AF729921

Basic Biostatistics

NIPA® GENX ELECTRONIC RESOURCES & SOLUTIONS P. LTD.
New Delhi-110 034

the Authors

Dr. Biswanath Patra, currently, Professor and Head in Animal Genetics & Breeding at MB Veterinary College, Dungarpur, Rajasthan. He has also worked as Associate Professor and Head in Animal Genetics & Breeding at RPS Veterinary College, Mahendragarh, affiliated with LUVAS, Hissar. He has more than twelve years of working experience with various aspects in Molecular Genetics, Pharmacology, Immunogenetics, Bioinformatics and Biostatistics in USA with several high impact NCBI publications including nature group journal. He has also teaching experiences in Animal Genetics & Breeding and Biostatistics. He did Master's in Genetics and Animal Breeding from GBPUA&T, Pantnagar and a Ph.D. in Animal Breeding from Indian Veterinary Research Institute (IVRI).

He did Postdoctoral Research in several world famous institutes including UT South-western Medical Centre at Dallas, the home of seven noble laureates till date. He learned genome Bioinformatics at Thomas Jefferson University, Philadelphia. He has given several awards wining oral and poster presentations from his research works in national and international symposiums. For several years he has presented his research works at highly reputed Experimental Biology meetings in USA. E-Email:patrabn1@gmail.com, biswanathpatra1@gmail.com, Phone no. +91-8777806047(M).

Dr. Bharat Bhushan, PhD, ARS 1985, Principal Scientist & Head Division of Animal Genetics, ICAR-Indian Veterinary Research Institute (ICAR-IVRI), Izatnagar – 243 122 (UP), India, did his PG and PhD in Animal Breeding from G. B. Pant University of Agric. & Tech., Pantnagar, India. He was selected for Agricultural Research Service (ARS) in the year 1985 and joined as Scientist in the discipline of Animal Genetics & Breeding at ICAR. He had more than 38 years of experience in conducting research on various aspects of Animal Genetics, Animal Breeding, Molecular Genetics and Genomics. Published more than 261 research papers in journals of national and international repute, co-authored 9 lead/review papers, 150 invited lectures and presented more than 350 research papers in various national and international conferences/symposia. He was involved in guidance of more than 100 research scholars for their MVSc and PhD degrees (30 as Chairman: 16 MVSc; 14 PhD). He had undergone the Advanced training on "Genetic resistance to diseases", Washington State University, Pullman, USA. Recipient of several national awards/recognition including the Best Teacher Award 2006-07 of ICAR-IVRI and Best Paper Award for highest Impact Factor by ICAR-IVRI. As Program Coordinator, successfully organized an International

Basic Biostatistics

NIPA® GENX ELECTRONIC RESOURCES & SOLUTIONS P. LTD.
New Delhi-110 034

About the Authors

Dr. Biswanath Patra, currently, Professor and Head in Animal Genetics & Breeding at MB Veterinary College, Dungarpur, Rajasthan. He has also worked as Associate Professor and Head in Animal Genetics & Breeding at RPS Veterinary College, Mahendragarh, affiliated with LUVAS, Hissar. He has more than twelve years of working experience with various aspects in Molecular Genetics, Pharmacology, Immunogenetics, Bioinformatics and Biostatistics in USA with several high impact NCBI publications including nature group journal. He has also teaching experiences in Animal Genetics & Breeding and Biostatistics. He did Master's in Genetics and Animal Breeding from GBPUA&T, Pantnagar and a Ph.D. in Animal Breeding from Indian Veterinary Research Institute (IVRI).

He did Postdoctoral Research in several world famous institutes including UT South-western Medical Centre at Dallas, the home of seven noble laureates till date. He learned genome Bioinformatics at Thomas Jefferson University, Philadelphia. He has given several awards wining oral and poster presentations from his research works in national and international symposiums. For several years he has presented his research works at highly reputed Experimental Biology meetings in USA. E-Email:patrabn1@gmail.com, biswanathpatra1@gmail.com, Phone no. +91-8777806047(M).

Dr. Bharat Bhushan, PhD, ARS 1985, Principal Scientist & Head Division of Animal Genetics, ICAR-Indian Veterinary Research Institute (ICAR-IVRI), Izatnagar – 243 122 (UP), India, did his PG and PhD in Animal Breeding from G. B. Pant University of Agric. & Tech., Pantnagar, India. He was selected for Agricultural Research Service (ARS) in the year 1985 and joined as Scientist in the discipline of Animal Genetics & Breeding at ICAR. He had more than 38 years of experience in conducting research on various aspects of Animal Genetics, Animal Breeding, Molecular Genetics and Genomics. Published more than 261 research papers in journals of national and international repute, co-authored 9 lead/review papers, 150 invited lectures and presented more than 350 research papers in various national and international conferences/symposia. He was involved in guidance of more than 100 research scholars for their MVSc and PhD degrees (30 as Chairman: 16 MVSc; 14 PhD). He had undergone the Advanced training on "Genetic resistance to diseases", Washington State University, Pullman, USA. Recipient of several national awards/recognition including the Best Teacher Award 2006-07 of ICAR-IVRI and Best Paper Award for highest Impact Factor by ICAR-IVRI. As Program Coordinator, successfully organized an International

Training for Afghan Scientists sponsored by Ministry of External Affairs, Government of India, New Delhi. Patented a portable large animal restraining device "TRAVIPORT" Design No. 255486. Evaluated a number of Masters and PhD thesis received from various Institutions/Universities of India. Also evaluated a PhD thesis received from Charles Sturt University, Locked Bag 588, Bocrooma St., Wagga Wagga, NSW, Australia. (+91-9411607873 (Mobile); Fax: +91-581-230 3284;E- Email: bhushan.drbharat@gmail.com; bharat.bhushan1@icar.gov.in

Dr. Hitesh Purohit is presently working as an Assistant Professor in the Department of Animal Genetics and Breeding at Bhumika College of Veterinary Sciences and Research Centre, Sigra, Mahendragarh which is affiliated from LUVAS, Hisar. Dr. Hitesh Purohit has done MVSc and PhD in Animal Genetics and Breeding. Dr. Hitesh Purohit received MVSc from Bihar Veterinary Science University, Patna in the year 2019 and PhD degree from Rajasthan Veterinary and Animal Sciences University, Bikaner in the year 2023. Dr. Hitesh Purohit's also worked as Assistant Professor in AGB at RPS Veterinary College, Mahendragarh, Haryana. Dr. Hitesh Purohit is working with Shri Godham Mahatirtha Pathmeda Lokpunyarth Nyas, the world's largest Gausewa organization, Chief Advisor (Gochikitsa and Gosamvardhan) as a honouree service.

Dr Parth Gaur is currently pursuing PhD in Animal Genetics and Breeding from ICAR-NDRI, Karnal. He has done his MVSc from Lala Lajpat Rai University, Hissar, Haryana. He is also a recipient of ICAR JRF SRF fellowship and has published research articles and popular articles in various national and international journals. He has also served as Assistant Professor at RPS Veterinary College, Mahendragarh, affiliated with LUVAS, Hissar.

Basic Biostatistics

Biswanath Patra
Postdoctoral Fellow, USA
Professor and Head
Animal Genetics & Breeding
MB Veterinary College, Dungarpur, Rajasthan-314 001

Bharat Bhushan
Principal Scientist & Head
Division of Animal Genetics
ICAR-Indian Veterinary Research Institute
Izatnagar, Bareilly, Uttar Pradesh-243 122

Hitesh Purohit
Assistant Professor
Animal Genetics and Breeding
Bhumika College of Veterinary Sciences and Research Centre
Sigra, Mahendragarh, Haryana-123 029

Parth Gaur
Department of Animal Genetics and Breeding
ICAR- National Dairy Research Institute
Karnal, Hariyana-132 001

NIPA® GENX ELECTRONIC RESOURCES & SOLUTIONS P. LTD.
New Delhi-110 034

NIPA® GENX ELECTRONIC RESOURCES & SOLUTIONS P. LTD.

101,103, Vikas Surya Plaza, CU Block
L.S.C.Market, Pitam Pura, New Delhi-110 034
Ph : +91 11 4386 0225, 9717133558, 9540816132
E-mail: newindiapublishingagency@gmail.com
Website: www.niparesources.com

© 2025, Publisher

Print ISBN: 978-93-58874-65-5

ebook ISBN: 978-93-58879-08-7

All rights reserved. No part of this publication may be reproduced, stored in a retrieval system or transmitted in any form or by any means, including electronic, mechanical, photocopying recording or otherwise without the prior written permission of the publisher or the copyright holder.

This book contains information obtained from authentic and highly reliable sources. Reasonable efforts have been made to publish reliable data and information, but the author/s, editor/s and publisher cannot assume responsibility for the validity, accuracy or completeness of all materials or information published herein or the consequences of their use. The work is published with the understanding that the publisher and author/s are not attempting to render any professional services. The author/s, editor/s and publisher have attempted to trace and acknowledge the copyright holders of all material reproduced in this publication and apologize to copyright holders if permission and/or acknowledgements to publish in this form have not been taken. If any copyrighted material has not been acknowledged, please write to us and let us know so that we may rectify the error, in subsequent reprints.

Trademark Notice: NIPA®, the NIPA® logos and their presentations (the way they are written/presented) in this book are the trademarks of the publisher and hence may not be used without written permission, if copied or used without authorization, the infringer will be prosecuted as per law.

NIPA® also publishes books in a variety of electronic formats. Some content that appears in print may not be available in electronic books, and vice versa

Composed and Designed by NIPA®.

Acknowledgement

My sincere thanks to Prof. Vinod Kumar Sharma, Hon'ble Dean, MB Veterinary College, Dungarpur, Rajasthan, India for encouraging me to write the book on "Basic Biostatistics" for UG and PG students as well as academicians according to the Veterinary Council of India Syllabus. My personal thanks to the management authorities namely Mr. Ashish Anand, Director and Mr. Kapil Anand, Secretary of this college for appointing me as Professor, Animal Genetics and Breeding for motivating and writing this book for the benefit of the college students and faculties.

I am very much grateful to my parents Mr. Ashutosh Patra (a retired High School Science teacher) who taught me mathematics in my school life and my mother Mrs. Basanti Patra for their blessings and encouragement in my academic pursuits, which helped me to reach at this level. The gratitude of beloved Gurudev and Lord Shiva (MY PRABHU) doesn't require any thanks, because the outcome of this book in this form is possible because of their blessings. I am also highly beholden to Pratima Kuity Patra, Shri Dulal Kuity and Mrs. Gauri Kuity for their continuous encouragement of pursuing my academic careers.

I express my gratitude to each and everyone who have helped me to publish this book. I also extend my special thanks to Indranil Samanta, Devi Prasad Isore, Uttam Sarkar, Parth Gaur, Hitesh Purohit, Shashikant Choudhury and Sandeep Gera for their motivation in my professional career. I am highly indebted to my research supervisors namely Dr R.K.S. Bais, Principal Scientist, ICAR-Central Avian Research Institute (CARI), Izatnagar, Bareilly, Uttar Pradesh\, India and Dr Bharat Bhushan, Principal Scientist & Head, Division of Animal Genetics, ICAR-Indian Veterinary Research Institute (IVRI), Izatnagar, Bareilly, Uttar Pradesh, India, who have guided me during MVSc and PhD degree programmes, respectively.

Above all my sincere thanks to the publishing agency "New India Publishing Agency (NIPA)" New Delhi, India for designing and publishing this book and making my dream successful.

Authors

Preface

The main aim for writing the book on **"Basic Biostatistics" is** to assist the students and faculties of the biological sciences in teaching and research. Besides, this book provides the basic concepts to apply appropriate experimental designs and statistical tools for recording and analysis of data for better inferences. The book chapters in this book cover basic concepts required for teaching at undergraduate as well as postgraduate levels keeping in view Indian and foreign universities/institutions. Statistical methods applied to biological sciences are known as Biostatistics or Biometrics, which have originated from agricultural and biomedical research.

The peculiar characteristics that distinguish biometrics within statistics is the fact that biological measurements are variable, not only because of measurement error, but also from their natural variability from genetic and environmental sources. These sources of variability must be taken into account when drawing conclusions about any biological materials. Accounting for these sources of variation has led to the development of experimental designs that incorporate blocking, covariates and repeated measures. Appropriate techniques for analysis of data from these designs and others are discussed properly in this book.

The syllabus of BVSc & AH and BSc (Ag.) has been covered in 15 chapters as per Veterinary Council of India (VCI). After this, two chapters have been written for PG and PhD students of Animal Genetics and Breeding disciplines, which contain Descriptive Biostatistics, Inferential Biostatistics and Linear Models in animal breeding. The last chapter covers the special topics question banks for UG and PG students, which will help the students to practice and prepare for various university examinations, ICAR examinations like JRF, SRF, ARS and CSIR etc. All the Veterinary Universities/Colleges and Agriculture Colleges in India and also Animal Science Colleges over the globe teach Basic Biostatistics to Animal Genetics and Breeding students in UG, PG and PhD level with the aim to learn and develop new methodology for livestock improvement, so that their productivity could be enhanced to the optimum level.

This book has attractive features that are simple language, easy to understand, compiled and arranged properly with illustrations etc. Besides, statistical theory has been illustrated with examples for better understanding of all statistical principles. Solved numerical problems have also been discussed as reference for better understanding. In the last chapter, model question papers have been discussed and answers have also been provided to help the students. Two chapters written on linear models will be very helpful for PG and PhD students. The authors have provided a lot of good information in a comprehensive manner in this book, which will not only help the students to better understand the subject but also help for the preparation of various competitive examinations.

The help rendered by MB Veterinary College, Dungarpur, Rajasthan, India for publishing this book is duly acknowledged. The authors welcome the necessary suggestions from concerned faculties and students for improving this book in future editions.

Authors

Contents

1

Introduction

Statistics

It is a branch of mathematics, which comprises several areas like Algebra, Trigonometry, Calculus, Geometry, Statistics and Probability etc.

Biostatistics is a mathematical science which analyses numerical facts or biometric data of biological events following certain rules like design of experiments, analysis of data, presentation of data with tables, figures hypothesis testing, making a probability prediction and conclusion etc.

or it is an art of analysis and presentation of data following certain rules and international norms accepted by international journals for pharmacology experiments, animal farm data, human and animal medicines data, epidemiology data, non-parametric data etc.

What is statistics?

The number of introductory or elementary texts on the **subject of statistics indicates how important the subject has become for everyone in the biological sciences**. However, the fact that there are many texts might also suggest that we have yet to discover a fool proof method of presenting what is required. The problem confronted in **biological statistics** is as follows:

When there is a need to make a set of numerical observations in biology, it has been observed that the values are scattered. Then questions arise to know whether these values differ because of factors e.g. treatments or these are part of a 'background' i.e., natural variation. You need to evaluate what the numbers actually mean, and to represent them in a way that readily communicates their meaning to others.

The subject of statistics grips

A. The design of study should be in order to reveal most information efficiently.

B. Collection of data.

C. Analysis of data.

D. The presentation of suitably summarized information, often in a graphical or tabular form.

E. The interpretation of the analyses in a manner that communicates the findings accurately globally.

Strictly, this broad numerical approach to biology is correctly termed '**biometry**' but we shall adopt the more generally used term '**statistics**' to cover all aspects. Statistics (meaning this entire process) has become one of the essential tools in modern biology.

Definition of Biostatistics

Biostatistics is a branch of Mathematics which deals with the collection, compilation, presentation, analysis and interpretation of numerical data or quantitative information of biological organism.

Applications of Statistics

Industry or commerce	**Education**
Economics	Medical science
Biology	Agriculture
Botany	Veterinary
Astronomy	Dairy
Physics	Sociology
Chemistry	Meteorology, etc.

Functions of Statistics

I. It presents facts in a definite form.

II. It simplifies the mass of figures.

III. It facilitates comparison.

IV. It helps in formulating and testing hypotheses.

V. It helps in probability and prediction.

VI. It helps in the formulation of suitable policies.

Statistical packages for analysis using a computer

Followings are more frequently used statistics software for data analysis:

1. Statistical Package for the Social Sciences (SPSS), (IBM SPSS Version 20 (www-01.ibm.com/software/analytics/spss, accessed 9 October 2012))
2. Statistical Analysis System (SAS) is a command-driven software package used for statistical analysis and data visualization. (https://www.sas.com/en_in/software/stat.html)
3. Stata (Stata 12, StataCorp, 2011, Stata Statistical Software: Release 12. College Station, TX: StataCorp LP (www.stata.com/products, accessed 9 October 2012)).
4. GraphPad Prism & Biostatistics Training

   GraphpadPrisam SnapGene - Molecular Biology & Cloning , SnapGene Software (graphstats.net).
5. The R Project for Statistical Computing

   R is a free software environment for statistical computing and graphics. It compiles and runs on a wide variety of UNIX platforms, Windows and MacOS. To download R, please choose your preferred CRAN mirror.

Origin of statistics

Statistics is not a new discipline but as old as human society itself. It has been used right from the existence of life on this earth. The word 'statistics' has come from the Italian word 'statesman' or German word 'statistik' which means State. The term statistics was first used by Professor Gottfried Achenwall (1719 –1772) to refer to the subject matter as a whole. Achenwall defined statistics as "the political science of the several countries". The word statistics appeared for the first time in the famous book, Elements of Universal Erudition by Baran J. F. Von Bielfeld and translated by W. Hooper in 1770.

- The present form of statistical methods is based on the theory of probability which marked a major step in the history of the world.
- The gamblers of 17th century attracted the attention of De Moivre, Galileo and great mathematicians like James Bernoulli, Daniel Bernoulli, Laplace and Karl Gaus who discovered and developed the theory of probability while estimating the chance of winning or losing in gamble.

- **Abraham De Moivre (1667 – 1754)** French mathematician known for de Moivre's formula, a formula that links complex numbers and trigonometry, and for his work on the **normal distribution and probability theory**.
- Laplace and Gauss independently arrived at the same results of De Moivre.

Great mathematician **Jaques Quetelet** (1796 – 1874) introduced statistical methods to the social sciences. He also founded the science of anthropometry and developed the body mass index (BMI) scale, originally called the Quetelet Index.

Sir Francis Galton (1822 – 1911) developed the concept of Regression, is called the **father of Biostatistics.**

Karl Pearson (1857 – 1936) developed the Chi-square test of goodness of fit & Correlation.

Sir Ronald A Fisher (1890 – 1962) made major contributions in the field of Experimental Designs and Analysis of Variance, is called the **father of Statistics**.

William Sealy Gosset (13 June 1876 – 16 October 1937), Gosset published under the pen name Student and developed most famously **Student's t-distribution** – originally called Student's "z" – and "**Student's test of statistical significance**".

Statistics in Veterinary and Animal Science

The **published scientific literature** is full of studies in which statistical procedures are employed. Look in any of the relevant scientific journals and notice the number of times reference is made to mean ± SEM (standard error of mean), to statistical significance, to *P*-values or to *t*-tests or Chi-squared analysis or analysis of variance or multiple regression analysis.

- In the animal health sciences, there are an increasing number of independent **diagnostic services** that will **analyze samples for the benefit of health monitoring** and maintenance.
- The **pharmaceutical and agrochemical industries** are required to demonstrate both the safety and the efficacy of their products in an indisputable manner. Such data invariably require a statistical approach to establish and illustrate the basis of the claim for both these aspects.

The experts need to be aware of the appropriate statistical procedures in order to play their proper roles.

The subject of **epidemiology**, there is an essential need for you to understand the types of trials and investigations that are carried out and to know the meaning of the terms associated with them.

Under the term **evidence based veterinary medicine** (EBVM) – by which we mean the conscientious, explicit and judicious use of current best evidence to inform clinical judgements and decision-making in veterinary care (Cockcroft and Holmes, 2003)

Types of variable

A **variable** is a characteristic that can take values which *vary* from individual to individual or group to group, e.g. height, weight, litter size, blood count, enzyme activity, coat colour, percentage of the flock which are pregnant, etc. Clearly some of these are more readily quantifiable than others

1. **Categorical (qualitative) variable** – an individual belongs to any one of two or more distinct categories for this variable. A *binary* or a dichotomous variable is a particular type of categorical variable defined by only *two* categories; for example, pregnant or non-pregnant, male or female.

    - **Nominal scale** – the distinct categories that define the variable are unordered and each can be assigned a name, e.g. coat colors (piebald, roan or grey).

    - **Ordinal scale** – the categories that constitute the variable have some intrinsic order; for example, body condition scores, **subjective intensity of fluorescence of cells in the fluorescence microscope.**

2. **Numerical (quantitative) variable** – consisting of numerical values on a well-defined scale, which may be:

    - **Discrete (discontinuous) scale**, i.e. data can take only particular integer values, typically many traits of bioeconomic importance in animal production (survival, calving difficulty, disease resistance) present discontinuous distributions of phenotypes.

    - **Continuous scale**, for which all values are theoretically possible (perhaps limited by an upper and/or lower boundary), e.g. height, weight, speed, concentration of a chemical constituent of the blood or urine.

Biological causes of variation

- Genetics/Breeds – e.g. greater variability in the whole cattle population compared with just Friesians.
- Environment/Locations – e.g. body weight varies with **feeding, housing intercurrent disease**, etc.
- Gender – sexual dimorphism is common.
- Age – many biological data are influenced by **age and maturity, e.g. the quantity of body fat.**

Technical errors

Measurement errors are due to factors which are, typically, **human** (e.g. variations within and between observers) or **instrumental**, but may also be attributed to differences in conditions (e.g. different laboratories).

A **systematic error** is one in which the observed values have a tendency to be above (or below) the true value; the result is then said to be ***biased***. When the observed values are evenly distributed above and below the true value, **random errors**, due to unexplained sources, are said to be occurring.

Instrumental errors

- **Instrumental errors** arise for a number of reasons. Providing we are aware of the potential problem, the causes are often correctable or reducible.
- With a ***systematic offset* or *zero error*, a 'blank' sample** consistently reads other than zero. It is common in colorimetry and radioisotope measurements.
- ***Non-linearity* is a systematic error**, commonly seen in the performance of strain gauges, thermocouples and colorimeters.
- ***Proportional* or *scale error*** is usually due to electronic gain being incorrectly adjusted or altered after calibration; it results in a systematic error.
- **Hysteresis is a systematic error** commonly encountered in measurements involving galvanometers. It may require a standard measurement procedure, e.g. always adjusting input down to desired level.

- **Instability or drift** – electronic gain calibration may drift with temperature and humidity giving rise to an intermittent but systematic error, resulting in an unstable baseline.
- **Random errors** are commonly seen in attempts to measure with a sensitivity beyond the limits of resolution of an instrument. Most instruments carry a specification of their accuracy, for example it is no use attempting to measure to the nearest gram with a balance accurate only to 10 g.

This is the **essence of quality control** and is of paramount importance in a diagnostic laboratory. Quality control in the laboratory is about ensuring that processes and procedures are carried out in a consistently satisfactory manner.

Terms relating to measurement quality

Two terms which are of major importance in understanding the principles of biological measurement are **precision** and **accuracy**. It is essential they are understood early in a consideration of the nature of data measurement.

- **Precision** refers to **how well repeated observations agree with one another.**
- **Accuracy** refers to how well the **observed value agrees with the true value.**

Several other terms, all of which describe aspects of reliability, are in use and these are defined as follows:

- **Repeatability** is concerned with gauging the similarity of replicate, often duplicate, measurements of a particular technique or instrument or observer under identical conditions, e.g. measurements made by the same observer in the same laboratory.
- **Reproducibility** (sometimes called **method agreement**) is concerned with determining how well two or more approaches to measuring the same quantity agree with one another, e.g. measurements made by the same observer but using different methods

Populations and samples

The population may be either a **real (or finite) group** or a **hypothetical (or infinite) group**. For example, if we are interested in the growth rate of pigs in Suffolk, then the population is all pigs in Suffolk. This is a real or finite population.

A population includes all representatives of a particular group, whereas **a sample is a subgroup drawn from the population**. We aim to choose a sufficiently large sample in such a manner that it is representative (i.e. is typical) of the population.

Random sampling and random allocation

In order for the sample to be representative, strictly, there should be **random selection** from all possible members of the entire population, implying that the individuals should be selected using a method based on chance.

However, in reality, random selection is generally not feasible (for example, in an observational study or in a clinical study when the disease under investigation is rare).

Types of statistical procedures

Statistical procedures can be divided into descriptive statistics and inferential statistics.

- **Descriptive statistics**. We use these techniques to reduce a data set to manageable proportions, summarizing the trends and tendencies within it, in order to represent the results clearly. From these procedures we can produce **diagrams, tables and numerical descriptors**. Numerical descriptors include measures that convey where the **center of the data set lies, like the arithmetic mean or median, and measures of the scatter or dispersion of the data, such as the variance or range**.
- **Inferential statistics**. Statistical inference is the process of generalizing from the **sample to the population**: it enables us **hypothesis testing to draw conclusions about certain features of the population** when only a subset of it, the sample, is available for investigation. One aspect of inferential statistics is the **estimation** of **population parameters using sample data.** A parameter, such as the mean or proportion, describes a particular feature of the **distribution of a variable in the entire.**

In that case, it is important that we try to ensure that the individuals in the **sample are a true reflection of those in the population of interest**, and that, if groups are to be compared, we check that the **individuals in the different groups are comparable with similar baseline characteristics.**

Note that for allocating individuals into treatment groups in an experimental situation, principles of **random allocation (randomization)** should also be employed to avoid subjective influence and ensure that the groups are comparable.

Greek letters used in Biostatistics

In the subject of biostatistics, **Greek letters** play a significant role. These symbols are used to represent various constants, special functions, and variables related to statistical concepts. Let's explore some of the key Greek letters and their applications:

Alpha (α): Often used as the significance level in hypothesis testing. It represents the probability of making a Type I error (rejecting a true null hypothesis).

Beta (β): Represents the probability of a Type II error (failing to reject a false null hypothesis). It is also used in power calculations.

Gamma (γ): Appears in the context of **gamma distributions and gamma functions**. These are commonly used in modeling waiting times and reliability analysis.

Delta (δ): Used to denote **differences or changes in a variable**. For instance the delta method is employed to approximate the variance of a function of random variables.

Epsilon (ε): Represents **the error term in regression models**. It signifies the deviation between the observed data and the model's predictions.

Pi (π): Famously associated with the circle's circumference and area. In statistics, it is used in formulas related to **probability distributions, such as the normal distribution**.

Sigma (σ): Denotes the **standard deviation of a population** or sample. It also appears in summation notation (Σ) for adding up a series of values.

Sigma^2 (σ^2): Population variance is symbolized by σ^2.

Mu (μ): Represents the **population mean**. It is commonly used in statistical formulas and equations.

Theta (θ): Appears in parameter **estimation and maximum likelihood estimation**. It represents unknown parameters in statistical models.

Omega (ω): Often used to denote **angular velocity or frequency**. In statistics it can represent the population variance.

Remember that the use of **Greek letters** in biostatistics is widespread, and each letter carries specific meanings within the context of statistical analysis.

Definition of Primary Data

What is data in statistics?

Data are individual pieces of factual information mostly numerical, recorded and used for the purpose of analysis. It is the raw information from which statistics are created. Statistics are the results of data analysis, its interpretation and presentation.

In statistics, data refers to a collection of discrete or continuous values that convey information. It is a systematic record of a specific quantity, and the diverse values of that quantity together represent the data set. Data is used for the purpose of the analysis process, and statistics are the result of data analysis.

Primary data set is originated for the first time by the researcher through direct efforts and experience, specifically for the purpose of addressing his research problem. It is also known as the first hand or raw data. Primary data collection is quite expensive, as the research is conducted by the organization or agency itself, which requires resources like investment and manpower. The data collection is under direct control and supervision of the investigator.

The data can be collected through various methods like surveys, observations physical testing, mailed questionnaires, questionnaire filled and sent by enumerators, personal interviews, telephonic interviews, focus groups, case studies, etc.

Definition of secondary data

Secondary data implies **second-hand information which is already collected and recorded by any person other than the user for a purpose, not relating to the current research problem**. It is the readily available form of data collected from various sources like censuses, government publications internal records of the organization, reports, books, journal articles, websites and so on.

Secondary data offer several advantages as it is easily available, **saves time and cost to the researcher**. But there are some disadvantages associated with this, as the data is gathered for the purposes other than the problem in mind, so the usefulness of the data may be limited in a number of ways like relevance and accuracy.

Basis For Comparison	Primary Data	Secondary Data
Meaning	Primary data refers to the first-hand data gathered by the researcher himself.	Secondary data means data collected by someone else earlier.
Data	Real time data	Past data
Process	Very involved	Quick and easy
Source	Surveys, observations, experiments, questionnaire, personal interview, etc.	Government publications, websites, books, journal articles, internal records etc.
Cost effectiveness	Expensive	Economical
Collection time	Long	Short
Specific	Always specific to the researcher's needs.	May or may not be specific to the researcher's need.
Available in	Crude form	Refined form
Accuracy and Reliability	More	Relatively less

Important terms

The data collected by the investigator himself or by his agent from the sample or population are called as the **primary data.**

The data collected from the available sources is known as **secondary data.**

Width or length of the Class Interval is the difference between the upper boundary and lower boundary of the same class.

Lower and upper limits of class intervals are known as **Class Limits.**

Class frequency or frequency is the number of observations in that class.

A frequency distribution in statistics is a representation that displays the number of observations within a given interval.

Statistics- It is branch of science which deals with collection, classification, tabulation, presentation, analysis and interpretation of data. (Father of Statistics – R.A. Fisher).

Biostatistics is the application of statistical methods to the problems of biology including human biology, medicine and public health (Father of Biostatistics - Francis Galton).

Variable is any property of an individual (person, tree, rat etc.) that can be expressed in numerical terms.

Attributes: Qualitative characteristics of an individual which shows variability is known as attributes like colour, number of legs, weight, speed, etc.

Constant: It is a particular type of variable which does not vary from other member of a group to another or within a particular set of defined conditions.

Parameter: A statistical measure pertaining to a population is called a parameter. A parameter is any numerical property descriptive of population. (e.g.) mean, standard deviation of the population. **It is denoted by Greek letter e.g. μ for mean, σ^2 for variance and σ for standard deviation etc**.

Continuous variable

If a variable takes an intermediate value between any specified interval, it is called a continuous variable. e.g. the weight an animal.

Discrete or discontinuous variable

If a variable takes only integral values, then it is called a discrete or discontinuous variable. e.g. number of animals in a farm.

Conclusion

Data are individual pieces of factual information mostly numerical, recorded and used for the purpose of analysis. It is the raw information from which statistics are created. Statistics are the results of data analysis, its interpretation and presentation.

- The best motivation for wrestling with statistical concepts is the need to know the meaning of a data set of your own.
- Remember that statistical procedures cannot enhance poor data. Providing the data has been acquired with sufficient care and in sufficient numbers.
- The statistical procedures can supply you with sound summary statements and interpretative guidelines, hypothesis testing and drawing conclusions.
- Biostatistics is the art of presenting biological data, its demonstration in table or figures, analyses and drawing conclusions about sample and population.

2

Classification and Tabulation of Data

2.1 What is a frequency distribution?

A **frequency distribution** shows the frequencies of occurrence of the observations in a data set. Often the distribution of the observed data is called an empirical frequency distribution, in contrast to the theoretical probability distribution determined from a mathematical model.

The **class interval** is a term used in statistics when we are given a continuous series. Class means a group of numbers in which items are placed such as 0-10, 10-20, 20-30, etc. Class interval refers to the numerical width of any class in a particular distribution.

2.2 Relative frequency distributions

Although creating a frequency distribution is a useful way of describing a set of observations, it is difficult to compare two or more frequency distributions if the total number of observations in each distribution is different. A way of overcoming this difficulty is to calculate the proportion or percentage of observations in each class or category. These are called **relative frequencies** and each is obtained by dividing the frequency for that category by the total number of observations (column 3 of Table 2.1). The sum of the relative frequencies of all the categories is unity (or 100%) apart from rounding errors.

Table 2.1: Frequency distribution of of 188 lambs birth weight data

Birth weight (kg.)	No. of lambs	Relative frequency	Relative frequency (%)	Cumulative frequency
1.50-1.75	7	0.037	3.72	3.72
1.75-2.00	24	0.128	12.77	16.49
2.00-2.25	32	0.170	17.02	33.51
2.25-2.50	40	0.213	21.28	54.79
2.50-2.75	25	0.133	13.30	68.09
2.75-3.00	20	0.106	10.64	78.72
3.00-3.25	18	0.096	9.57	88.30
3.25-3.50	22	0.117	11.70	100.00
Total	188	1.000	100.00	

2.3 Cumulative relative frequency distributions

We form a **cumulative relative frequency distribution** by adding the relative frequencies of individuals contained in each category and all lower categories, and repeating this process for each category. The cumulative relative frequencies are tabulated in column 5 of Table 2.1 and the distribution drawn in the **cumulative relative frequency polygon** of Figure 2.1 is termed as **Ogive**. e.g. The **Lorenz Curve** is a valuable tool used to analyse income distribution and wealth inequality. It is a cumulative percentage curve that visually represents the deviation of the actual distribution from the ideal line of equal distribution.

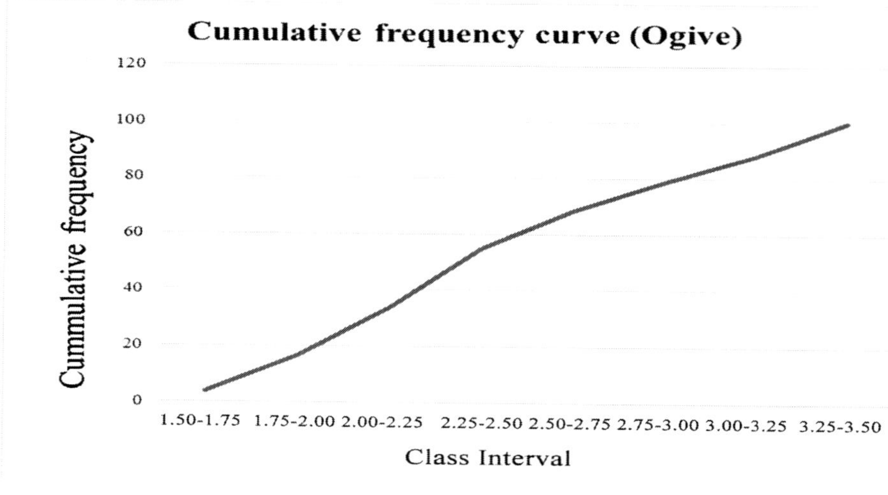

Fig. 2.1: Cumulative frequency distribution of 188 lambs birth weight data

We can evaluate the **percentiles** (often called *centiles*) of the frequency distribution from this cumulative frequency distribution. Percentiles are the values of the variable that divide the total frequency into 100 equal parts.

In particular, the 50th percentile (called the **median**) is the value of the variable that divides the distribution into two halves; 50% of the individuals have observations less than the median, and 50% of the individuals have observations greater than the median. Often the 25th and the 75th percentiles are quoted (these are called the **lower (first)quartile** and **upper (third) quartile**, respectively); 25% of the observations lie below the lower quartile and 25% of the observations lie above the upper quartile, the distance between these quartiles being the **interquartile range**. The 5th and 95th percentiles enclose the central 90% of the observations.

1. Tabulation of Data

It is the presentation of numerical data in row and column wise with appropriate information which will be easily understood for statistical analysis. There are different kind of tables as per many folds information on it.

Rules for well-constructed tables

I. Include a concise, informative and unambiguously defined title.

II. Give a brief heading for each row and column.

III. Include the units of measurement.

IV. Give the number of items on which any summary measure (e.g., a percentage) is based.

V. When providing a summary statistic (e.g., the mean) always include a measure of precision (e.g., a confidence interval).

VI. Give figures only to the degree of accuracy that is appropriate (as a guideline, one significant figure more than the raw data).

VII. Do not give too much information in a table.

VIII. Remember that it is easier to scan information down columns rather than across rows.

1. Simple tabulation: contain one set of data or it is one way classification.

Examples: Table 1. State wise population of livestock in different state

State	**Population**
Rajasthan	
Punjab	
Haryana	
Total	

2. Complex tabulation: contain many fold data, two way or more classifications.

i. Double or two way table: Such type of tables contains data classified on the basis of two attributes.

State	**Population**		**Total**
	Male	**Female**	
Rajasthan			
Punjab			
Haryana			
Total			

ii. Treble or three way table: classified basis on three attributes.

Example: Population of livestock in different states is divided according to sex and again subdivided according to age groups.

State	Population						Total		
	Male			Female					
	Young	Adult	Total	Young	Adult	Total	Young	Adult	Total
Rajasthan									
Punjab									
Haryana									
UP									
Total									

iii. Manifold table: Such type of tables contains data classified on the basis of more than three attributes.

Example: Population of livestock in different states is divided according to species again subdivided according to sex and further subdivided according the age groups.

State		Population						Total		
		Male			Female					
		Young	Adult	Total	Young	Adult	Total	Young	Adult	Total
Bihar	C									
	B									
	T									
Punjab	C									
	B									
	T									
Total										

Note: C: Cattle; B: Buffalo; T: Total

iv. Tabulation for frequency distribution.

To prepare the tables for frequency distribution two methods are used.

(i) **Exclusive methods**: In this the upper limit of a class and lower limit of the next higher class is same. The values equal to upper limit of a class interval is excluded from that class and included in the net higher class.

Age group (in months)	Number (frequency)
0-3	
3-6	
6-9	
9-12	

(ii) **Inclusive method:** In this, values equal to lower and upper limit of a class interval are included the same class.

Age group (in months)	Number (frequency)
0 –10	
11 – 20	
21 – 30	
31 - 40	

3. Diagrams

A diagram is a graphic representation of data and may take several forms. Diagrams are most useful when we want to convey information quickly, and they should serve as an adjunct to more formal statistical analysis.

Rules for well-constructed diagrams

I. Keep it simple and avoid unnecessary 'frills' (e.g. making a simple pie chart, histogram or bar chart three-dimensional).

II. Include a concise, informative and unambiguously defined title.

III. Label all axes, segments and bars, if necessary, using a legend or key showing the meaning of the different symbols used.

IV. Present the units, the numbers on which summary measures are based, and measures of variability where appropriate.

V. Avoid exaggerating the scale on an axis, perhaps by omitting the zero point, so as to distort the results.

VI. Include a break in the scale only if there is no other satisfactory way of demonstrating the extremes.

VII. Show coincident points in a scatter diagram.

VIII. Ensure that the method of display conveys all the relevant information (e.g. pairing).

1 Categorical (qualitative) data

When data are categorical or qualitative, then each observation belongs to one of a number of distinct categories or classes.

(a) Bar chart

A **bar chart** is a one dimensional diagram in which every category of the variable is represented; the length of each bar, which should be of constant width, depicts the number or percentage of individuals belonging to that category.

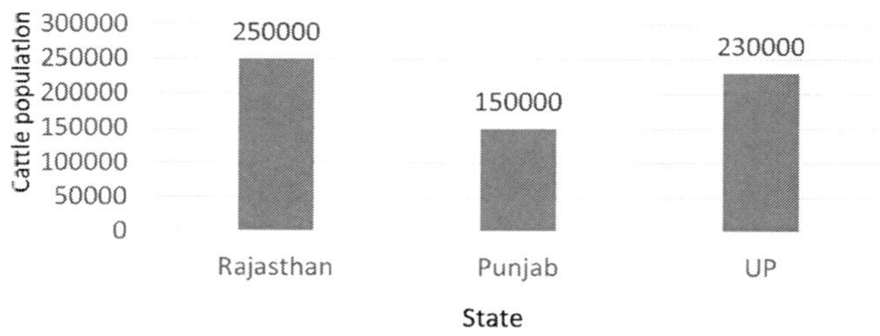

Fig. 3.1: Cattle population during the year 1990-1991

(b) Pie chart

A **pie chart** is a circle divided into segments with each segment portraying a different category of the qualitative variable (Figure 3.2). The total area of the circle represents the total frequency, and the area or angle of a given sector is proportional to the percentage of individuals falling into that category.

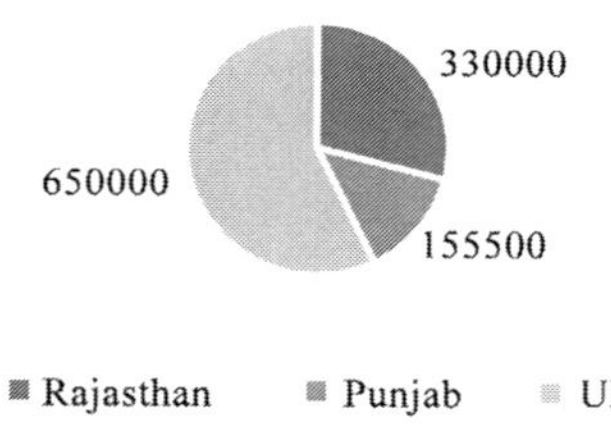

Fig. 3.2: Cattle population during the year 2000-2001

2. Numerical (quantitative) data

When the data are numerical or quantitative, we may show every data value, for example in a *dot diagram*, or we may display only a summary of the data, for example in a *histogram*.

(a) Dot diagram

If the data set is of a manageable size, the best way of displaying it is to show every value in a **dot diagram/plot**. When we investigate a single numerical variable, we can mark each observation as a dot on a line calibrated in the units of measurement of that variable, plotted horizontally or vertically.

- If the data are in a single group, the diagram will look like a horizontal line to the x axis.
- When we are comparing the observations in two or more groups, we can draw a dot diagram with the horizontal axis designating the groups, and the vertical axis representing the scale of measurement of the variable.

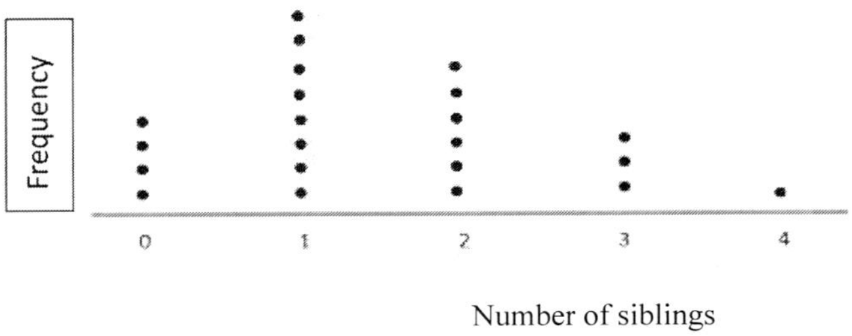

Fig. 3.3: Dot diagram of frequency for siblings

(b) Histogram

The frequency distribution of a quantitative variable can be displayed as a **histogram**. This is a two-dimensional diagram in which usually the horizontal axis represents the unit of measurement of the variable of interest, with each class interval being clearly delineated. The histogram gives a good picture of the frequency distribution of the variable. The distribution is **symmetrical** if its shape to the right of a central value is a mirror image of that to the left of the central value. The frequency distribution is **skewed to the right (positively skewed)** if the right-hand tail is extended, and **skewed to the left (negatively skewed)** if the left-hand tail is extended.

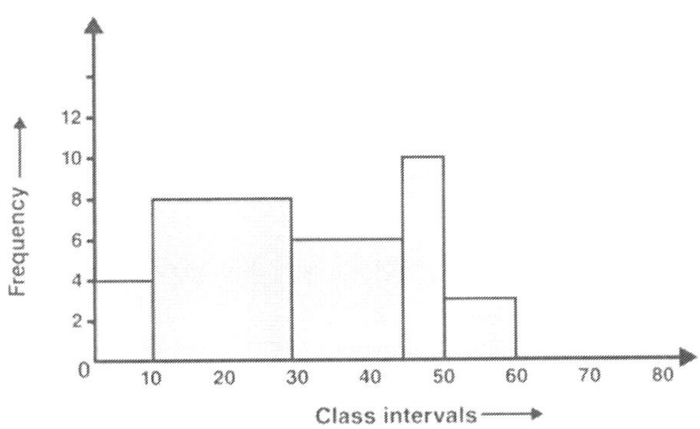

Fig. 3.4a: An asymmetric histogram

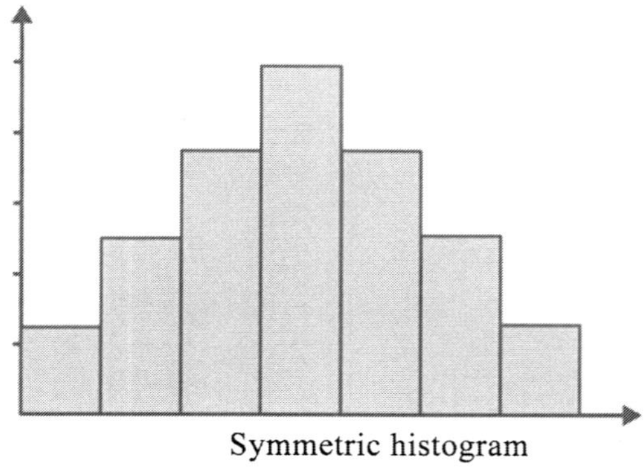

Fig. 3.4b: The symmetric histogram

(c) Stem-and-leaf diagram

We often see a mutation of the histogram, called a **stem-and-leaf diagram**, in computer outputs. Each vertical rectangle of the histogram is replaced by a row of numbers that represent the relevant observations. The stem is the core value of the observation (e.g. the unit value before the decimal place) and the leaves are represented by a sequence of ordered single digits, one for each observation, that follow the core value (e.g. the first decimal place). Plotting the data in this way provides an easily assimilated description of the distribution of the data whilst, at the same time, showing the raw data.

Fig. 3.5: A Stem-and-leaf diagram

(d) Box-and-whisker plot

Another diagram that we often see in computer outputs, the **box-and-whisker plot** (or **box plot**), provides a summary of the distribution of a data set. The scale of measurement of the variable is usually drawn vertically. The diagram comprises a box with horizontal limits defining the upper and lower quartiles and representing the interquartile range, enclosing the central 50% of the observations, with the median marked by a horizontal line within the box. The whiskers are vertical lines extending from the box as low as the 2.5th percentile and as high as the 97.5th percentile (sometimes the percentiles are replaced by the minimum and maximum values of the set of observations).

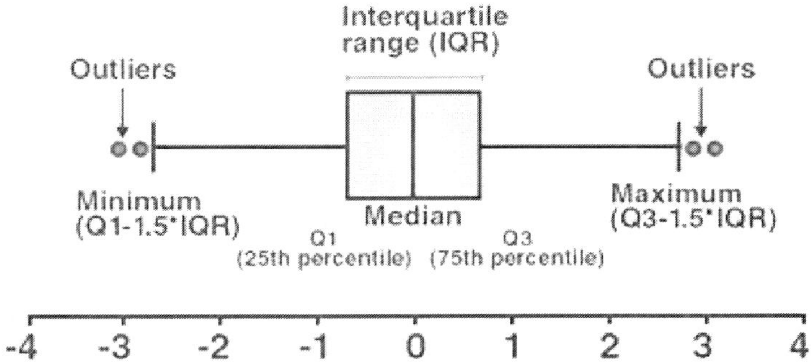

Fig. 3.6: A Box-and-whisker plot diagram

(e) Scatter diagram

The **scatter diagram** is an effective way of presenting data when we are interested in examining the relationship between two variables which may be numerical or ordinal. The diagram is a two-dimensional plot in which each axis represents the scale of measurement of one of the two variables. Using this rectangular coordinate system, we relate the value for an individual on the horizontal scale (the *abscissa*) to the corresponding value for that individual on the vertical scale (the *ordinate*) by marking the relevant point with an appropriate symbol. Coincident points should be identifiable. We can discern possible relationships between the variables by observing the scatter of points, and then we may join the points to produce a line graph.

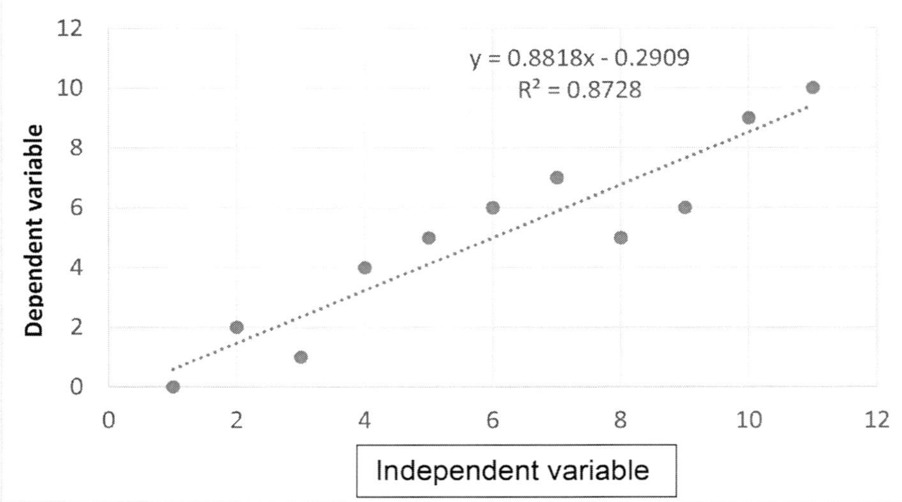

Fig. 3.7: Scatter diagram having independent variable on X axis and dependent variable on Y axis displaying equation of a straight line and value of R^2.

(f) Cumulative Frequency Curve (Ogive)

A curve that represents the cumulative frequency distribution of grouped data on a graph is called a Cumulative Frequency Curve or an Ogive. Representing cumulative frequency data on a graph is the most efficient way to understand the data and derive results.

There are two types of Cumulative Frequency Curves (or Ogives)

- More than type Cumulative Frequency Curve
- Less than type Cumulative Frequency Curve

The two types of ogives are less than ogive and greater than or more than ogive. In a less than ogive, the frequencies of all preceding classes are added to the frequency of a class. In a more than ogive, the frequencies of the succeeding classes are added to the frequency of a class.

How to plot a more than type ogive

1. In the graph, put the lower limit on the x-axis
2. Mark the cumulative frequency on the y-axis.
3. Plot the points (x,y) using lower limits (x) and their corresponding Cumulative frequency (y)

4. Join the points by a smooth freehand curve. It looks like an upside down *S*.

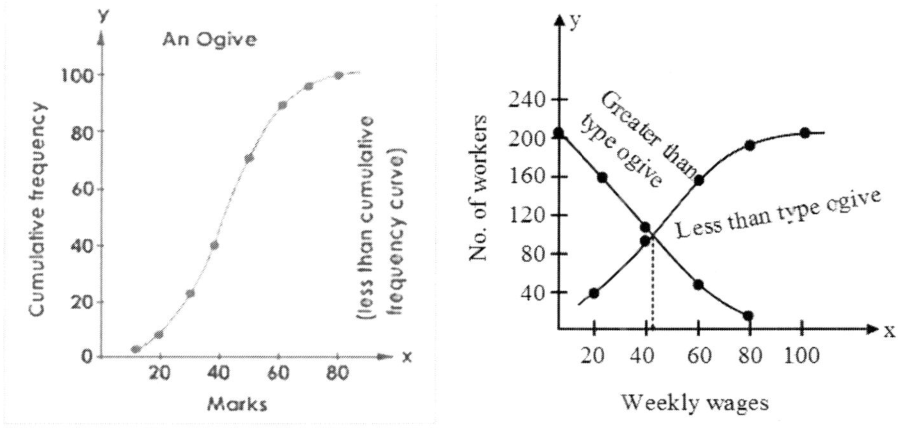

Fig. 3.8: Cumulative Frequency Curve (Ogive)

(g) Frequency Polygon

A frequency polygon is a graphical form of representation of data. It is used to depict the shape of the data and to depict trends. It is usually drawn with the help of a histogram but can be drawn without it as well. A histogram is a series of rectangular bars with no space between them and is used to represent frequency distributions.

In order to create a frequency polygon, one must follow these steps:

- Creation of a histogram.
- **Finding the midpoints for each bar that exists on the histogram.**
- Placing a point on the origin of the histogram and its end.
- Connection of the points.

Frequency polygons provide us with an understanding of the shape of the data and its trends.

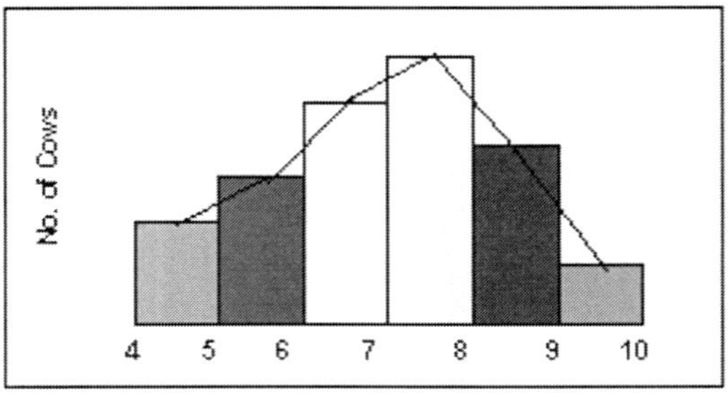

Fig. 3.9: Frequency Polygon

h) Frequency curve

It is a smooth free hand curve drawn through the vertices of the frequency polygon. They are enclosed by the frequency curve the same as that of the histogram or frequency polygon but its shape is not with sharp edges.

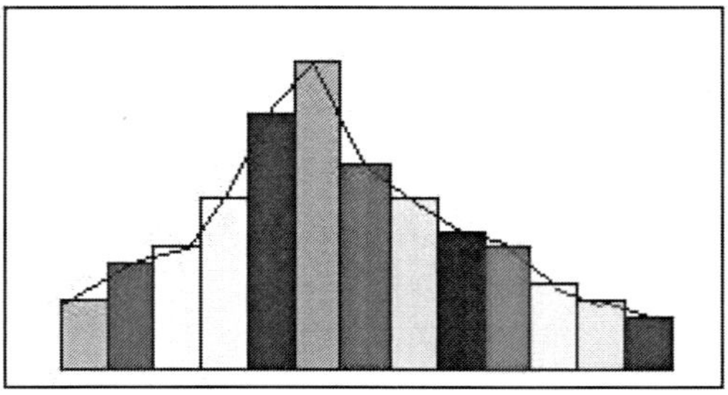

Fig. 3.10: Showing a Frequency curve

i) Pictogram

- Tabular data can also be represented by pictogram, cartogram, maps and pictures as these device help in attracting the attention to statistical matter which when presented in the ordinary diagrammatic form is very often ignored.
- **Pictograms** are diagrams of pictorial or semi-pictorial nature and are drawn in different sizes according to scale. Though they are useful in attracting the attention of the people, they very often check on tables, ignoring the pictorial diagrams.
- They cannot be made use of with certain complicated data.

Giraffe	
Zebra	
Tiger	
Wolf	
Bear	

Fig. 3.11: Pictogram of wild animals

j) Cartogram

A cartogram is a thematic map of a set of features (countries, provinces, etc.), in which their geographic size is altered to be directly proportional to a selected variable, such as travel time, population, or Gross National Product etc.

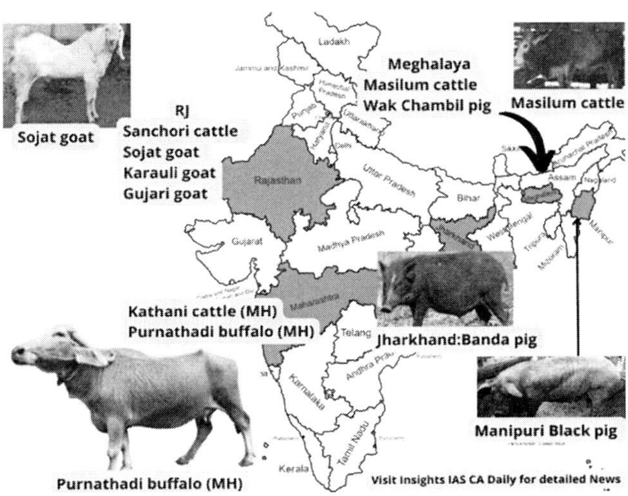

Fig. 3.12: Cartogram of 10 new indigenous breeds of livestock

Important takeaways from chapter

- Tabulation of data in statistics involves organizing and presenting data in a systematic and structured manner, typically using tables or charts. This process helps in summarizing and making data more understandable, which is crucial for analysis and decision-making.
- **Width or length of the Class Interval** is the difference between the upper boundary and lower boundary of the same class.
- Lower and upper limits of class interval are known as **Class Limits.**
- **Class frequency or frequency** is the number of observations in that class.
- **A frequency distribution** in statistics is a representation that displays the number of observations within a given interval.
- Tabulation of data is a fundamental step in data analysis and reporting. It enhances the clarity, comparability, and accessibility of data, making it an essential tool for decision-making, research, and communication in various fields.
- **Array Series:** Arrangement of data in ascending or descending order of their magnitude.
- **Cross-Tabulation (Contingency Table):** Display the relationship between two categorical variables by creating a table that shows how the categories of one variable correspond to the categories of another.
- **Bar Chart:** Use rectangular bars to represent data values for different categories or groups. Bar charts can be vertical (column chart) or horizontal.
- **Histogram:** Represent the distribution of continuous data by dividing it into intervals and using bars to display the frequency or density within each interval.
- **Pie Chart:** Show the composition of a whole by dividing it into segments (slices) that represent the relative proportions of different categories or components.
- **Line Graph:** Connect data points with lines to visualize trends and changes over a continuous range, typically used for time series data.
- **Scatter Plot:** Plot individual data points on a graph to explore relationships and correlations between two variables.

- **Box Plot (Box-and-Whisker Plot):** Display the distribution of a dataset's summary statistics, including median, quartiles, and potential outliers.
- **Heatmap:** Visualize the intensity of relationships between two categorical variables using color-coding.
- **Cartogram/Mapgraph:** Diagrams are made to represent geographical situations.
- **Historigram (Line graph):** is a time series graph.
- **Histogram:** Is a frequency distribution graph. It is used to calculate value of mode.
- **Ogive:** Is a cumulative frequency curve. It is used to the value of median.
- **Pictograms:** In statistics, **pictograms** are charts in which icons represent numbers to make it more interesting and easier to understand. A key is often included to indicate what each icon represents. All icons must be of the same size, but a fraction of an icon can be used to show the respective fraction of that amount.
- **Class mark** is a statistical term that refers to the average of the upper limit and the lower limit of a class in a frequency distribution. It is also known as the mid-value or the middle value of each class interval.

3

Measures of Central Tendency

The central tendency measure is defined as the number used to represent the center or middle of a set of data values. The three commonly used measures of central tendency are the mean, median, and mode.

Measures of location (averages)

The term **average** refers to any one of several measures of the **central tendency** of a data set.

(a) Arithmetic mean

The most commonly used measure of central tendency is the **arithmetic mean** (usually abbreviated to the **mean**). It is obtained by **adding together the observations in a data set and dividing by the number of observations in the set**. If the continuous variable of interest is denoted by x and there are n observations in the sample, then the sample mean (pronounced x bar) is

Arithmetic Mean

$$\text{Mean } \overline{X} = \sum_{i=1}^{n} \frac{x_i}{n} = \frac{x_1 + x_2 + \ldots.. + x_n}{n}$$

$$\text{Arithmetic Mean} = \frac{3+2+6+5}{4} = 4$$

Example, Find the arithmetic mean of the five numbers. Solution: First five numbers are 2, 3, 5, 7 and 11. The mean or the average formula is = (2 + 3 + 5 + 7 +11)/5 = 5.6

- The mean has the disadvantage that its value is **influenced by outliers**. An **outlier** is an observation whose value is highly inconsistent with the main body of the data. An outlier with an excessively large value will tend to increase the mean unduly, whilst a particularly small value will decrease it.
- The mean is an appropriate measure of central tendency if the distribution of the **data is symmetrical**.

The mean will be 'pulled' to the right (**increased in value**) if the distribution is skewed to the right, and 'pulled' to the left (**decreased in value**) if the distribution is skewed to the left.

- **The arithmetic mean and the median are close or equal in value if the distribution is symmetrical.**

Weighted Mean is an average computed by giving different weights to some of the individual values. If all the weights are equal, then the weighted mean is the same as the arithmetic mean.

Formula for weighted mean

The Weighted mean for a given set of non-negative data $x_1, x_2, x_3, \ldots\ldots x_n$

with non-negative weights $w_1, w_2, w_3, \ldots\ldots w_n$

can be derived by the formula given below:

Weighted mean = $(w_1x_1 + w_2x_2 + \ldots\ldots\ldots\ldots + w_nx_n) / (w_1 + w_2 + \ldots\ldots\ldots\ldots + w_n)$

Where,

Xi is the repeating value

Wi is the number of occurrences of Xi weight

$\bar{X}$ bar is the weighted mean

(b) Median (positional average)

Another frequently used measure of central tendency is the **median**. The median is the central value in the set of *n* observations which have been arranged in **rank order**, i.e. the observations are arranged in an increasing (or decreasing) order of magnitude. The median is the **middle value of the ordered set** with as many observations above it as below it. **The median is the 50th percentile.** In median, the middle most value of the series is taken as the representative value. Therefore, the median is a **positional average.**

Calculation of Median

- If n is odd then the median is found by starting with the smallest observation in the ordered data set then counting until (n+1)/2th observation is reached.
- If n is even then the median is average of two middle values.

Example

The weights (grams) of 19 male Hartley guinea pigs were:

314, 991, 789, 556, 412, 499, 350, 863, 455, 297, 598,510, 388, 642, 474, 333, 421, 685, 536

If we arrange the weights in rank order, they become:

297, 314, 333, 350, 388, 412, 421, 455, 474, **499**, 510,536, 556, 598, 642, 685, 789, 863, 991

The median, shown in bold, is the (19 + 1)/2 =10th weight in the ordered set; this is 499 g.

- **The advantage of the median is that it is not affected by outliers** or if the distribution of the data is skewed. Thus the median will be less than the mean if the data are skewed to the right, and greater than the mean if the data are skewed to the left.
- **A disadvantage of the median is that it does not incorporate all the observations** in its calculations, and it is difficult to handle mathematically.

(c) Geometric mean

If we take the logarithm (generally to base 10 or to base *e*) of each value of a data set which is skewed to the right, we usually find that the distribution of the log-transformed data becomes more symmetrical. In this case, the arithmetic mean of the log-transformed values is a useful measure of location. However, it has the disadvantage that it is measured on a log scale. We therefore, **convert it back to the original scale by taking its antilogarithm**; this is the **geometric mean**. The distribution of biological data, if not symmetrical, is frequently skewed to the right; we could then calculate the geometric mean to represent an average value.

In mathematics, the geometric mean is a mean or average which indicates a central tendency of a finite set of real numbers by using the product of their values (as opposed to the arithmetic mean which uses their sum). **The geometric mean is defined as the nth root of the product of n numbers, i.e., for a set of numbers a_1, a_2, ..., a_n, the geometric mean is defined as**

$$\left(\prod_{i=1}^{n} a_i\right)^{\frac{1}{2}} = n\sqrt{a_1 a_2 .... a_n}$$

or, equivalently, as the arithmetic mean in logscale:

$$\exp\left(\frac{1}{n}\sum_{i=1}^{n} In\, a_i\right)$$

Example,

For instance, the geometric mean of two numbers, say 2 and 8, is just the square root of their product, that is, 4

As another example, the geometric mean of the three numbers 4, 1, and 1/32 is the cube root of their product (1/8), which is 1/2.

The distribution is more symmetrical, and the geometric mean represents the central tendency of the log transformed data much.

Example,

This can be written as,

$$Log\, GM = \frac{1}{n}\log(x_1 \times x_2 \times \ldots x_n)$$

$$= \frac{1}{n}(\log x_1 + \log x_2 + \ldots. + \log x_n)$$

$$= \frac{\sum \log x_i}{n}$$

Therefore Geomatric Mean is,

$$GM = Anti\log \frac{\sum \log x_i}{n}$$

Where n = f1 + f2 +…..+ fn

It is also represented as,

$$G.M. = n\sqrt{\prod_{i=1}^{n} x_i}$$

For any group data GM can be written as,

$$GM = Anti\log \frac{\sum f \log x_i}{n}$$

Relation Between AM, GM and HM

To learn the relation between the AM, GM and HM, first we need to know the formulas of all these three types of the mean.

Assume that "x" and "y" are the two number and the number of values = 2, then

AM = (a+b)/2

⇒ 1/AM = 2/(a+b) (1)

GM = √(ab)

⇒ GM^2 = ab (2)

HM= 2/[(1/a) + (1/b)]

⇒ HM = 2/[(a+b)/ab

⇒ HM = 2ab/(a+b) (3)

Now, substitute (1) and (2) in (3), we get

HM = GM^2 /AM

⇒ GM^2 = AM × HM

Or else,

GM = √[AM × HM]

Hence, the relation between AM, GM and HM is GM^2 = AM × HM

Question: Find the geometric mean of the following grouped data for the frequency distribution of weights.

Weights of ear heads (g)	No. of ear heads (f)	Mid x	Log x	f log x
60-80	22	70	1.845	40.59
80-100	38	90	1.954	74.25
100-120	45	110	2.041	91.85
120-140	35	130	2.114	73.99
140-160	20	150	2.176	43.52
Total	160			324.2

From the given data, n = 160

We know that the G.M. for the grouped data is

$$GM = \text{Antilog}\frac{\sum f \log x_i}{n}$$

GM = Antilog (324.2 /160)

GM = Antilog (2.02625)

GM = 106.23

Therefore, the G.M = 106.23

Some of the important properties of the G.M. are

- **The G.M. for the given data set is always less than the arithmetic mean for the data set.**
- If each object in the data set is substituted by the G.M., then the product of the objects remains unchanged.
- The ratio of the corresponding observations of the G.M., in two series is equal to the ratio of their geometric means
- The products of the corresponding items of the G.M., in two series are equal to the product of their geometric mean.

Application of Geometric Mean

- It is used in stock indexes. Because many of the value line indexes which are used by financial departments use G.M.
- It is used to calculate the annual return on the portfolio.
- It is used in finance to find the average growth rates which are also referred to the compounded annual growth rate.
- **It is also used in studies like cell division and bacterial growth etc.**

(d) Harmonic mean

In mathematics, the harmonic mean is one of several kinds of average, and in particular, one of the Pythagorean means. It is sometimes appropriate for situations when the average rate is desired.

The harmonic mean can be expressed as the reciprocal of the arithmetic mean of the reciprocals of the given set of observations. As a simple example, the harmonic mean of 1, 4, and 4 is

$$\left(\frac{1^{-1}+4^{-1}+4^{-1}}{3}\right)^{-1}=\frac{3}{\frac{1}{1}+\frac{1}{4}+\frac{1}{4}}=\frac{3}{1.5}=2.$$

The harmonic mean H of the positive real numbers

$$H=\frac{n}{\frac{1}{x_1}+\frac{1}{x_2}+\ldots\frac{1}{x_n}}=\frac{n}{\sum_{i=1}^{n}\frac{1}{x_i}=\left(\frac{\sum_{i=1}^{n}x_i^{-1}}{n}\right)^{-1}}$$

1. Harmonic mean is a numerical average that is best used for fractions such as rates or ratios.
2. It **gives more weight to smaller values and less weight to larger values**.
3. It is often used in finance to average price multiples and in market analysis to identify patterns.
4. The harmonic mean is applied in the **finance to the average multiples like the price-earning ratio.**
5. It can be calculated for a series that includes negative values, but not for a series that includes a value of zero
6. The relationship between AM, GM, and HM is represented by the inequality AM $\geq$ GM $\geq$ HM.

Merits and Demerits of Harmonic Mean

The following are the merits of the harmonic mean:

- It is rigidly defined.
- It is based on all the views of a series, i.e. it cannot be computed by ignoring any item of a series.
- It is able to advance the algebraic method.
- It provides a more reliable result when the results to be achieved are the same for the various means adopted.
- It provides the highest weight to the smallest item of a series.
- It can also be measured when a series holds any negative value.
- It produces a skewed distribution of a normal one.
- It produces a curve straighter than that of the A.M. and G.M.

The demerits of the harmonic series are as follows

- The harmonic mean is greatly affected by the values of the extreme items
- It cannot be able to calculate if any of the **items is zero**
- The calculation of the harmonic mean is cumbersome, as it involves the calculation using the **reciprocals of the number**.

(e) Mode

1 A well-known but infrequently used measure of central tendency is the mode. It is the **most commonly occurring observation in a set of observations. The mode often has a different value from both the arithmetic mean and the median.** The modal group or modal class is the group or class into which most observations fall in a histogram.

2 In another context, we might use the mode to indicate the most common litter size in a breed of dogs, e.g. **the most common litter size of bearded collie dogs is seven.**

3 A distribution that has a **single mode or modal group is called unimodal**; a distribution that has two humps (i.e. modes or modal groups) separated by a trough is called **bimodal** even if the frequency of occurrence of the observations in the two modes or modal classes are not equal.

Example

Birth wt.	1.4	1.6	1.7	1.8	1.9
No. of kids	1	1	1	2	1

Here, birth weight 1.8 kg exhibited twice (2), hence the mode of the data is 1.8 kg.

Mode, Median, and Mean Formula

There exists an empirical relationship between mode, median and mean and this can be expressed using the formula:

Mode = 3 Median – 2 Mean

Merits and demerits of the mode as a measure of central tendency in statistics:

Merits (Advantages) of Mode

1. **Easy to calculate and understand**: Calculating the mode is straightforward. In some cases, it can be determined through simple observation or inspection. The concept of majority makes it easy to understand.

2. **Representative value:** The mode represents the value around which there is maximum concentration of observations. It serves as a good representative of the data.

3. **Not affected by extreme values**: Unlike the mean, the mode is not influenced by extreme values in the dataset. It can be calculated even if these extreme observations are unknown.

4. **No need for complete data**: Mode can be found even in open-ended frequency distributions. We only need the point of maximum frequency concentration; knowing all individual values is not necessary.

5. **Useful for both quantitative and qualitative data:** Mode can describe both quantitative (numeric) and qualitative (categorical) data. For example, it's used to measure taste preferences in surveys.

Demerits (Disadvantages) of Mode

1. **Not based on all observations:**: The mode considers only the highest concentration of frequencies and doesn't account for every item in the dataset.

2. **Indeterminate or ill-defined:**: Some distributions can be bimodal, trimodal, or multimodal, leading to multiple modes or no clear mode at all.

3. **Not rigidly defined:** There are different methods (inspection method and grouping method) to determine the mode, and they may yield different values.

4. **Affected by sampling fluctuations:** Mode is sensitive to sampling fluctuations, more so than the mean.

5. **Complex grouping process:** Calculating mode for unequal class intervals can be challenging.

Measure		**Formula**	
		Ungrouped data	**Grouped data**
Mean	**Arithmetic mean**	$\overline{X} = \frac{\sum_{i=1}^{n} Xi}{N}$	$\overline{X} = \frac{\sum fx}{\sum f}$
	Geometric mean	$\overline{X} = \sqrt[n]{X1 \times X2 \; .....Xn}$	$\overline{X} = \sqrt[N]{x_1^{f1} \times x_2^{f2} \times \; .....x_n^{fn}}$
	Harmonic mean	$\overline{X} = \frac{N}{\sum_{i=1}^{n} \frac{1}{Xi}}$	$\overline{X} = \frac{\sum fi}{\sum_{i=1}^{n} \frac{fi}{Xi}}$

1. Merits and demerits of Mean

Merits	**Demerits**
1. It is rigidly defined. 2. Easy to understand and calculate. 3. Based on all observations. **4. It is available for hypothesis testing and treatments.** 5. Arithmetic mean is **affected least by fluctuations of sampling.**	1. Cannot be determined by inspection or graphically. 2. Cannot be use if dealing qualitative characteristics. 3. Cannot be used in missing observations. 4. AM is affected by extreme values. **5. In asymmetrical distribution, AM is not a suitable measure of location.**

2. Merits and demerits of Median

Merits	Demerits
1. It is rigidly defined. 2. Easy to understandable and easy to calculate. 3. Based on all observations. 4. Median is **not affected by** extreme values..	1. In case an even number can not be determined exactly. 2. It is not based on all observations. 3. It **is not amenable for algebraic treatment**. 4. As compared to mean, it is **affected much by fluctuations of sampling**.

3. Merits and demerits of Mode

Merits	**Demerits**
1. Easy to calculate, like median, mode can be located in some cases merely by inspection. **2. Mode is not affected by extreme values**. 3. Mode can be conveniently located even if the **frequency distribution has class intervals of unequal magnitude.**	1. Mode is ill defined. It is not always possible to clearly find a mode. 2. A distribution may have two or more mode which are called bimodal and multimodal distribution. 3. It is not based on all observations. 4. It is not capable of further statistical treatment. 5. As compared with mean, mode is **affected to a greater extent by fluctuation of sampling.**

4

Calculation of Mean Median and Mode from Grouped Data

Mean of grouped data

The mean (average) of observations is the sum of the values of all the observations divided by the total number of observations. The mean of the data is given by $\overline{X} = (f_1 \times 1 + f_2 \times_2 + \ldots. + fnxn)/(f_1 + f_2 + \ldots + f_n)$. The mean Formula is given by,

Mean = ∑(fi.xi)/∑fi

Methods for calculating mean

Method 1: Direct method for calculating mean

Step 1: For each class, find the class mark x_i, as

x = 1/2(lower limit + upper limit)

Step 2: Calculate $f_i.x_i$ for each i.

Step 3: Use the formula **Mean = $\sum(f_i.x_i)/\sum f_{i.}$**

Example 1: Findout the mean of the following data.

Class interval	**0-10**	**10-20**	**20-30**	**30-40**	**40-50**
Frequency	12	16	6	7	9

Solution

We may prepare the table given below:

Class interval	Frequency f_i	Class Mark x_i	($f_i.x_i$)
0-10	12	5	60
10-20	16	15	240
20-30	6	25	150
30-40	7	35	245
40-50	9	45	405
	$\sum f_i$=50		$\sum f_i x_i$=1100

Mean = $\sum(f_i.x_i)/\sum f_i$ = **1100/50 = 22**

Method 2: Assumed – method for calculating mean

For calculating the mean in such a situation, we proceed as under.

Step 1: For each class interval, calculate the **class mark** x by using the formula: **x_i = 1/2 (lower limit + upper limit).**

Step 2: Choose a suitable value of mean and denote it by A. x in the middle as the assumed mean and denote it by A.

Step 3: Calculate the deviations $d_i = (x_i - A)$ for each i.

Step 4: Calculate the product $(f_i \cdot d_i)$ for each i.

Step 5: Find $n = \sum f_i$

Step 6: Calculate the mean, x, by using the formula: $\overline{X} = A + \sum f_i d_i / n$.

Example: Using the assumed-mean method, find the mean of the following data

Class interval	0-10	10-20	20-30	30-40	40-50
Frequency	7	8	12	13	10

Solution

Let A=25 be the assumed mean. Then we have,

Class interval	Frequency f_i	Mid value x_i	Deviation $d_i=(x_i-25)$	$(f_i.d_i)$
0-10	7	5	-20	-140
10-20	8	15	-10	-80
20-30	12	25=A	0	0
30-40	13	35	10	130
40-50	10	45	20	200
	$\sum f_i = 50$			$\sum(f_i.d_i) = 110$

Mean = $\overline{X} = A + \sum f_i d_i / n = (25 + 110/50) = 27.2$

Method 3: Step-Deviation method for calculating mean

When the values of x, and f are large, the calculation of the mean by the above methods becomes difficult. In such cases, we use the step-deviation method, given below.

Step 1: For each class interval, calculate the class mark x, where $\overline{X}$ **= 1/2 (lower limit + upper limit).**

Step 2: Choose a suitable value of x, in the middle of the x, column as the assumed mean and denote it by A.

Step 3: Calculate h = [(upper limit) – (lower limit)], which is the same for all the classes.

Step 4: Calculate $u_i = (x_i - A)/h$ for each class.

Step 5: Calculate fu for each class and hence find $\sum(f_i \times u_i)$.

Step 6: Calculate the mean by using the formula: $\overline{X} = A + \{h \times \sum(f_i \times u_i)/\sum f_i\}$

Example: Find the mean of the following frequency distribution

Class interval	50-70	70-90	90-110	110-130	130-150	150-170
Frequency	18	12	13	27	8	22

Solution

Prepare the table given below

Class interval	**Frequency f_i**	**Mid Value x_i**	**ui=(x_i-A)/h =(x_i-100)/20**	**($f_i.u_i$)**
50-70	18	60	-2	-36
70-90	12	80	-1	-12
90-110	13	100=A	0	0
110-130	27	120	1	27
130-150	8	140	2	16
150-170	22	160	3	66
	$\sum f_i = 100$			$\sum(f_i.u_i) = 61$

A = 100, h = 20, $\sum f_i = 100$ and $\sum(f_i . u_i) = 61$

$\overline{X} = A + \{h \times \sum(f_i . u_i) / \sum f_i\}$

$= 100 + \{20 \times 61/100\} = (100 + 12.2) = 112.2$

How to find the Median of grouped data?

In a grouped data, it is not possible to find the median for the given observation by looking at the cumulative frequencies. The middle value of the given data will be in some class interval. So, it is necessary to find the value inside the class interval that divides the whole distribution into two halves. In this scenario, we have to find the median class.

To find the median class, we have to find the cumulative frequencies of all the classes and n/2. After that, locate the class whose cumulative frequency is greater than (nearest to) n/2. The class is called the median class.

After finding the median class, use the below formula to find the median value.

$$Median = l + \left(\frac{\frac{n}{2} - cf}{f} \right) \times h$$

Where,

l is the lower limit of the median class

n is the number of observations

f is the frequency of median class

h is the class size

cf is the cumulative frequency of class preceding the median class.

Now, let us understand how to find the median of a grouped data using the formula with the help of an example.

Median of grouped data

Example: Following data represents the survey regarding the heights (in cm) of 51 girls of Class x. Find the median height.

Height (in cm)	Number of girls
Less than 140	4
Less than 145	11
Less than 150	29
Less than 155	40
Less than 160	46
Less than 165	51

Solution

To find the median height, first, we need to find the class intervals and their corresponding frequencies.

The given distribution is in the form of being less than type,145, 150 …and 165 gives the upper limit. Thus, the class should be below 140, 140-145, 145-150, 150-155, 155-160 and 160-165.

From the given distribution, it is observed that, 4 girls are below 140. Therefore, the frequency of class intervals below 140 is 4.

11 girls are there with heights less than 145, and 4 girls with height less than 140

Hence, the frequency distribution for the class interval 140-145 = 11-4 = 7

Likewise, the frequency of 145 -150= 29 – 11 = 18

Frequency of 150-155 = 40-29 = 11

Frequency of 155 – 160 = 46-40 = 6

Frequency of 160-165 = 51-46 = 5

Therefore, the frequency distribution table along with the cumulative frequencies are given below:

Solution

To find the median height, first, we need to find the class intervals and their corresponding frequencies.

The given distribution is in the form of being less than type,145, 150 ...and 165 gives the upper limit. Thus, the class should be below 140, 140-145, 145-150, 150-155, 155-160 and 160-165.

From the given distribution, it is observed that 4 girls are below 140. Therefore, the frequency of class intervals below 140 is 4. 11 girls are there with heights less than 145, and 4 girls with height less than 140

Hence, the frequency distribution for the class interval 140-145 = 11-4 = 7

Likewise, the frequency of 145 -150= 29 – 11 = 18

Frequency of 150-155 = 40-29 = 11

Frequency of 155 – 160 = 46-40 = 6

Frequency of 160-165 = 51-46 = 5

Therefore, the frequency distribution table along with the cumulative frequencies are given below:

Class intervals	Frequency	Cumulative frequency
Below 140	4	4
140 – 145	7	11
145 – 150	18	29
150 – 155	11	40
155 – 160	6	46
160 – 165	5	51

Here, n= 51.

Therefore, n/2 = 51/2 = 25.5

Thus, the observations lie between the class interval 145-150, which is called the median class.

Therefore,

Lower class limit = 145

Class size, h = 5

Frequency of the median class, f = 18

Cumulative frequency of the class preceding the median class, cf = 11.

We know that the formula to find the median of the grouped data is:

$$Median = l + \left(\frac{\frac{n}{2} - cf}{f} \right) \times h$$

Now, substituting the values in the formula, we get

$$Median = 145 + \left(\frac{25.5 - 11}{18} \right) \times 5$$

Median = 145 + (72.5/18)

Median = 145 + 4.03

Median = 149.03.

Therefore, the median height for the given data is 149. 03 cm.

How to find the mode of grouped data?

In the case of grouped data, it is not possible to identify the mode of the data, by looking at the frequency of data. In this scenario, we can determine the mode value by locating the class with the maximum frequency called modal class. Inside a modal class, we can locate the mode value of the data by using the formula:

$$Mode = l + \left(\frac{f_1 - f_0}{2f_1 - f_0 - f_2} \right) \times h$$

Where,

f_1 is the frequency of the modal class

f_0 is the frequency of the class preceding the modal class

f_2 is the frequency of the class succeeding the modal class

h is the size of the class intervals

l is the lower limit of the modal class

Now, let us understand how to find the mode of grouped data using this formula with the help of an example.

Mode of Grouped Data Example

Example 1

Example 1: A survey has been conducted by a group of students on 20 households in a locality as shown in the following frequency distribution table. Find the mode for the given data.

Size of family	1-3	3-5	5-7	7-9	9-11
No. of families	7	8	2	2	1

Solution

From the given table, it is observed that the maximum class frequency is 8, and the corresponding class interval is 3-5.

Therefore, the modal class for the given data is 3-5.

The lower limit of modal class, l = 3

Class size, h = 2

Frequency of modal class, $f_1 = 8$

Frequency of class proceeding to modal class, $f_0 = 7$

Frequency of class succeeding to modal class, $f_2 = 2$

We know that the formula to find the mode of the grouped data is:

$$Mode = l + \left(\frac{f_1 - f_0}{2f_1 - f_0 - f_2}\right) \times h$$

Now, substituting the values in the mode formula, we get:

$$Mode = 3 + \left(\frac{8 - 7}{2(8) - 7 - 2}\right) \times 2$$

Mode = 3 + (2/7)

Mode = (21+2)/7

Mode = 23/7

Mode = 3.286.

Therefore, the mode of the given grouped data is 3.286.

Example 2: Find the mean, mode, and median of the following data,

Class interval	**0-10**	**10-20**	**20-30**	**30-40**	**40-50**	**Total**
Frequency	8	16	36	34	6	100

Solution

We have,

Class interval	**Mid value x_i**	**Frequency f_i**	**Cumulative frequency**	**$f_i . x_i$**
0-10	5	8	8	40
10-20	15	16	24	240
20-30	25	36	60	900
30-40	35	34	94	1190
40-50	45	6	100	270
		$\sum f_i=100$		$\sum f_i . x_i=2640$

Mean = $\sum(f_i . x_i)/\sum f$

= 2640/100

= 26.4

Here, N = 100 ⇒ N / 2 = 50.

Cumulative frequency just greater than 50 is 60 and corresponding class is 20-30.

Thus, the median class is 20-30.

Hence, l = 20, h = 10, f = 36, c = c. f. of preceding class = 24 and N/2=50

Median, M_e = l + h{(N/2 – c_f)/f}

= 20+10{(50-24)/36}

Median = 27.2.

Mode = 3(median) – 2(mean) = (3 × 27.2) – (2 × 26.4) = 28.8.

5

Measures of Dispersion

What is dispersion?

In statistics, dispersion (also called variability, scatter, or spread) is the extent to which a distribution is stretched or squeezed. Common examples of measures of statistical dispersion are the variance, standard deviation, and interquartile range. For instance, when the variance of data in a set is large, the data is widely scattered. On the other hand, when the variance is small, the data in the set is clustered.

Dispersion is contrasted with location or central tendency, and together they are the most used properties of distributions.

Measure of dispersion

(a) Range

The **range** is defined as the difference between the largest and smallest observations. Range(X)= Max(X)- Min (X)

Example

The range in statistics for a given data set is the difference between the highest and lowest values. For example, if the given data set is {2,5,8,10,3}, then the range will be $10 - 2 = 8$. Thus, the range could also be defined as the difference between the highest observation and lowest observation.

- The range is an **easily determined measure of dispersion** of the observations of a numerical variable.
- It gives undue weight to extreme values and will, therefore, overestimate the dispersion of most of the observations if outliers are present.
- The range tends to increase in value as the number of observations in the sample increases.

(b) Interquartile range

The **interquartile range** is the range of values that encloses the central 50% of the observations if the observations are arranged in order of magnitude. **It is defined as the difference between the first and third quartiles.**

- The interquartile range is influenced neither by the presence of outliers nor by the sample size.

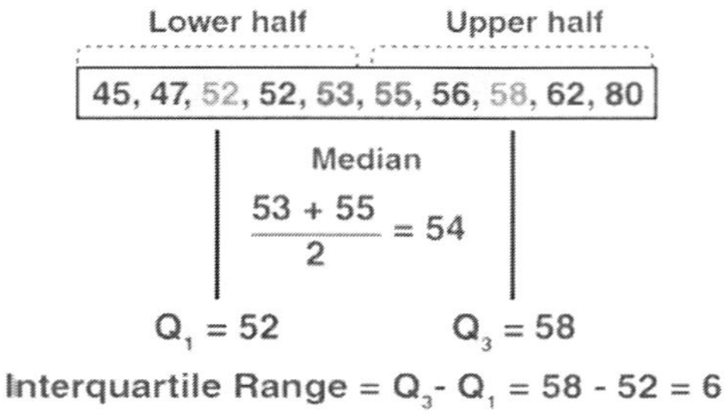

- It suffers from the disadvantage, in common with the range, of ignoring most of the observations as it is calculated from just two of them.
- Often the 25th and the 75th percentiles are quoted (these are called the **lower (first) quartile** and **upper (third) quartile**, respectively); 25% of the observations lie below the lower quartile and 25% of the observations lie above the upper quartile, the distance between these quartiles being the **interquartile range**.

c) Quartile deviation

The quartile deviation is less sensitive to extreme values because it only considers the middle 50% of the data. It is a robust measure of variation and is useful when the data set contains outliers. In contrast, **standard deviation is affected** by outliers because it's sensitive to extremely large or small values.

The quartile deviation can be defined mathematically as half of the difference between the upper and lower quartile. Here, quartile deviation can be represented as QD; Q_3 denotes the upper quartile and Q_1 indicates the lower quartile. Quartile deviation is also known as the Semi Interquartile range.

$$QD = \frac{1}{2}(Q_3 - Q_1)$$

$$Q_1 = l_1 + [i(\frac{N}{4} - cf)]/f$$

$$Q_3 = l_2 + [i(\frac{3N}{4} - cf)]/f$$

Where,

l_1 & l_2, are the lower limits of 1st and 3rd quartile class.

i = class interval

N = total number of observations

cf = Cumulative frequency just preceding the quartile class,

f = frequency of the quartile class.

Example

Calculate quartile deviation from the following data:

Wages of labour (Rs. per week)	**Frequency**	**Cumulative frequency**
Less than 35	14	14
35 – 37	62	76
38 – 40	99	175
41 – 43	18	193
Above 43	07	200

Answer:

$$Q.D. = \frac{Q3-Q1}{2},$$

for Q1 the $\frac{N}{4}$ th item = $\frac{200}{4}$ = 50th item.

$$Q_1 = l_1 + [i(\frac{N}{4} - cf)]/f$$

l = 35, N = 200, c.f. = 14, f = 62, I = 2.

$$Q_1 = 35 + \frac{\left(\frac{200}{\ } \quad 14\right)}{62} \text{ x } 2 = 36.16$$

$$Q_3 = l_2 + [i(\frac{3N}{4} - cf)]/f$$

For Q3 the 3N/4th item 3x200/4 = 150th

l = 38, N = 200, c.f. = 76, f = 99, i = 2

$$Q_3 = 38 + \frac{\frac{(3x200}{4} - 76)}{99} \text{ x } 2 = 38 + 1.49 = 39.49$$

QD = (39.49 – 36.16)/2 = 1.67

d) Variance

The arithmetic mean of squared deviations is called the variance. It is defined as mean sum of the square deviation of the values when values are expressed as deviation from the population mean.

Variance formula

Population	**Sample**
$\sigma^2 = \frac{\sum (x_i - \mu)^2}{N}$	$s^2 = \frac{\sum (x_i - \overline{x})^2}{n-1}$
x_i = elements in population	x_i = elements in sample
μ = population mean	$\overline{x}$ = sample mean
N = population size	n = sample size

In fact, when we select a sample of *n* observations from our population, we divide the sum of the squared deviations in the sample by *n* – 1 instead of *n*. It can be shown that this produces a better estimate (i.e. *unbiased*) of the population variance.

- Sample variance is denoted as s^2 & Population variance is denoted as σ^2 (sigma square).
- $S^2 = \frac{[\Sigma xi^2 - (\Sigma x)^2/N]}{n-1}$
- Degree of freedom for sample variance (s^2) is (n-1) whereas in case of population, DF is 'N'.

- **Unit of variance is the square of the unit in original data**

Importance of variance:

1. One of the most important measures of dispersion:
   i) Additivity or addition
   ii) Sub divisibility or partitioning
2. Based on all the observations of a set of data.
3. Easy to understand & easy to compute and commonly use used for most of the experiments.

Calculation of variance through direct formula : Population
$\sigma^2 = 1/N \sum (x_i - \overline{x})^2$

Sl.No.	Xi	$(Xi - \bar{X})$	$(Xi - \bar{X})$	$(Xi - \bar{X})2$
1.	2	2 – 4	-2	4
2.	3	3 – 4	-1	1
3.	4	4 – 4	0	0
4.	5	5 – 4	1	1
5.	6	6 - 4	2	4
Total	20		0	10
Mean $(\bar{X})$	4			

Sample variance $s^2 = [\sum x^2 - (\sum x)^2/n] / (n-1)$

Variance $s^2 = [\ 90 - 20^2 / 5\ ] / 4 = [90 – 400/5]/4$

$= (90 – 80)/4 = 10/4 = 2.5$

Standard deviation (SD) = $\sqrt{2.5} = 1.58$

e) ***Standard deviation (SD)***

It shows how much variation there is from the average (mean). A low SD indicates that the data points tend to be close to the mean, whereas a high SD indicates that the data are spread out over a large range of values.

- Standard deviation: $+\sqrt{\text{Variance}}$ *or* σ

  $SD = +\sqrt{[\sum x^2 - (\sum x)^2/N] / (N\text{-}1)}$.

  Some important points:

  - The concept of Standard deviation was given by Karl Pearson in 1893.
  - Standard deviation is the **best measure of dispersion.**
  - It is based on each and every item of the distribution.
  - We also know that the square root of any quantity is always positive or zero. Hence, the standard deviation is always non-negative.

- The SD uses all the observations in the data set.
- The SD is a measure of spread whose dimensionality is the same as that of the original observations, i.e. it is measured in the same units as the observations.
- The SD is of greatest use in relation to a symmetrically distributed data set that follows the Gaussian or Normal distribution. In this case, it can be shown that the interval defined by the (mean ± 2 SD) enccmpasses

the central 95% (approximately) of the observations in the population. In the example, the interval is 4.57 ± 2(0.319), i.e. from 3.93 to 5.21 mmol/l.

- For data that are Normally distributed, four times the standard deviation gives us an indication of the *range* of the majority of the values in the population. In the plasma potassium example, this is 4 × 0.319 = 1.28 mmol/l.

f) Coefficient of variation

Sometimes the standard deviation is expressed as a percentage of the mean; we call this measure the **coefficient of variation (CV).** It is a dimensionless quantity that can be used for comparing relative amounts of variation. However, these comparisons are entirely subjective because its theoretical properties are complex, so we do not recommend its use.

$$CV = \frac{\sigma}{\mu} \times 100\%$$

σ = population standard deviation

μ = population mean

Coefficient of variation is a **relative measure of dispersion** that is used to determine the variability of data. It is expressed as a ratio of the standard deviation to the mean multiplied by 100. It is a dimensionless quantity. As it is expressed in percentage, so it has no unit.

g) The standard error

The standard error of a statistic is the standard deviation of its sampling distribution or an estimate of that standard deviation. If the statistic is the sample mean, it is called the standard error of the mean.

$$SE = \frac{\sigma}{\sqrt{n}}$$

SE = standard error of the population

σ = population standard deviation

The standard error of the mean, or simply standard error, indicates how different the population mean is likely to be from a sample mean. It tells you how much the sample mean would vary if you were to repeat a study using new samples from within a single population.

Mean Absolute Deviation

- Value of mean deviation zero in a given data set.
- Average absolute deviation of the collected data set is the average of absolute deviations from a centre point of the data set.
- Mean absolute deviation is best used as it is more accurate and easy to use in real-life situations. The formula for mean absolute deviation (MAD) is as follows:

$$MAD = \sum_{i=1}^{n} \frac{|x_i - \overline{x}|}{n}$$

Where

x_i = Input data values

$\overline{X}$ = Mean value for a given set of data,

n = Number of data values

To find MAD, you need to following steps

- Calculate the mean for the given set of data.
- Find the difference between each value present in the data set and the mean that gives you the absolute value.
- Find the average of all the absolute values of the difference between the data set and the mean that gives the mean absolute deviation (MAD).

Now set the following table for MAD

Xi	Xi- $\overline{x}$	Absolute value of Xi- $\overline{x}$
26	-8	8
46	12	12
56	22	22
45	11	11
19	-15	15
22	-12	12
24	-10	10

Mean = (26 + 46 + 56 + 45 + 19 + 22 + 24)/7 = 238/7 = 34 i.e. $\overline{X}$ = 34

Now, let's find out the average of all the absolute values:

(8+12+22+11+15+12+10)/7=90/7=12.857)

Therefore, the mean absolute deviation of the given data set is 12.857.

Important takeaways from chapter

- **Statistical dispersion** means the extent to which numerical data is likely to vary about an average value i.e. how data is distributed.
- Absolute Measure of Dispersion-expresses the variations in terms of the **average of deviations of observations** and contains the **same unit** as the original data set. These are:
  - **Range:** difference between the maximum value and the minimum value given in a data set. Example: 1, 3,5, 6, 7 => Range = 7 -1= 6
  - **Variance:** The **average** of the **squared differences** from the **mean**. Variance (σ^2) = $\sum(X_i - \mu)^2/N$
  - **Standard deviation:** The square root of the average of the squared differences from the mean (Root mean square). The square root of the variance is known as the standard deviation i.e. S.D. = $\sqrt{\text{Variance}}$= (σ). $\sigma = \sqrt{\sum(Xi - \mu)^2/N}$
  - **Quartiles and quartile deviation:** The quartiles are values that divide a list of numbers into quarters. The quartile deviation is half of the distance between the third and the first quartile.
  - **Mean deviation:** Arithmetic mean of the absolute deviations of the observations from a measure of central tendency is known as the mean deviation (also called mean absolute deviation).
- Relative Measure of Dispersion used to **compare the distribution of two or more data sets** and this measure compares values **without units.**
  - Coefficient of Variation- C.V = σ/Mean *100

6

Moments, Skewness and Kurtosis

Moments

In statistics, moments are measures of the shape and variability of a data set. They are used to describe the location and dispersion of the data. There are several types of moments that can be calculated, each providing different information about the data set.

Moments is a term which refers to a measure of force with respect to its tendency which provides rotation. The strength of the tendency depends upon the amount of force and distance from the point of origin where force is exerted.

If a number of force is f_1, f_2, f_3........ f_n at distance of x_1, x_2, x_3.....x_n are applied, the moment of the first force about origin is f_1x_1, the moment of second force is f_2x_2 and so on.

So total moments of all the forces about origin, if total moment is divided by the total force, which is called moment actually arithmetic mean for frequency distribution.

Hence, arithmetic mean is called first moment about the origin

- **Arithmetic mean of various powers expressed as deviation from mean in any distribution is called the moment of the distribution.**
- **Central moments:** the moment about the mean is called the central moment. The rth moment of n observation about the mean X or central moment usually denoted by μ^r is defined as:

  $\mu_r=\Sigma(X-\bar{X})^r/N$

  For raw data where X1, X2, X3.......Xn are the 'n' observations and $\bar{X} = \Sigma(X)/n$
- The mean of the first power of deviation is the first moment about the mean. $\mu_1=\Sigma(X-\bar{X})^1/N$
- Since the sum of deviations of observation from arithmetic mean is always zero, the first moment μ_1 is always zero.

- The second moment about the mean μ_2 equal to **variance** as follow: $\mu_2=\Sigma(X-\bar{X})^2/N$

Third moment about the mean μ_3 define as Skewness as follow:

$\mu_3=\Sigma(X-\bar{X})^3/N$

The moments can be extended to higher power in the similar way but generally first four or up to μ_4 (fourth moment) define as Kurtosis is calculated:

$\mu_4=\Sigma(X-\bar{X})^4/N$

Measures of the shape of a distribution

The measures of the shape of a distribution are the coefficients of skewness and kurtosis.

Skewness is the degree of asymmetry of a distribution. Skewness (sk) is a measure of asymmetry of a frequency distribution. It shows if deviations from the mean are larger on one side than the other side of the distribution. If the population mean (μ) is known, then skewness is:

$$sk=\frac{n}{(n-1)(n-2)}\sum_i\left(\frac{y_i-\mu}{s}\right)^3$$

If the population mean is unknown, the sample mean (y) is substituted for μ and skewness is:

$$sk=\frac{n}{(n-1)(n-2)}\sum_i\left(\frac{y_i-\bar{y}}{s}\right)^3$$

For a symmetric distribution skewness is equal to zero. It is positive when the right tail is longer, and negative when the left tail is longer.

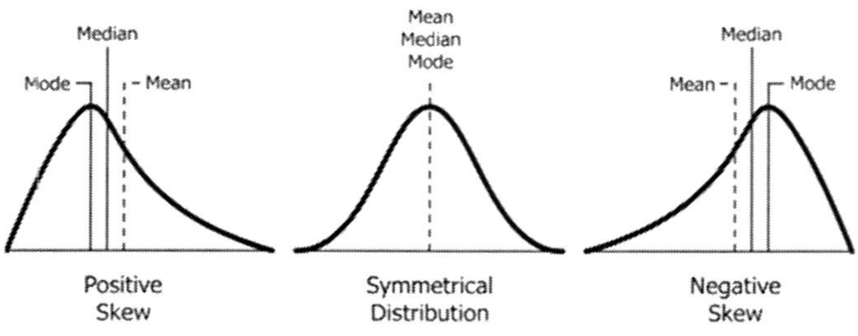

- In positively skewed distribution the mean is always greater than median and the median is always greater than the mode.
- In a negatively skewed distribution the mean is always lesser than median and the median is always lesser than the mode.

Parameter of peakedness

Kurtosis (Ku) is a measure of relative peakedness of a distribution. It is a shape parameter that characterizes the degree of peakedness. Kurtosis is a statistical measure that quantifies the shape of a probability distribution. It provides information about the tails and peakedness of the distribution compared to a normal distribution.

Tails: Tails are the tapering ends on either side of a distribution. They represent the probability or frequency of values that are extremely high or low compared to the mean. In other words, **tails represent how often outliers occur.**

Peakedness: Contrary to popular belief, kurtosis does not directly measure the "peakedness" of a distribution. Instead, it focuses on the shape of the tails. A distribution can be sharply peaked with low kurtosis, and a distribution can have a lower peak with high kurtosis. Thus, kurtosis measures "tailedness," not "peakedness"

Kurtosis (kt) **is a measure of flatness or steepness of a distribution**, or a measure of the **heaviness of the tails of a distribution**. If the population mean (μ) is known, kurtosis is:

$$kt = \frac{1}{n}\sum_i \left(\frac{y_i - \mu}{s}\right)^4 - 3$$

If the population mean is unknown, the sample mean ($\bar{y}$) is used instead and kurtosis is

$$kt = \frac{n(n+1)}{(n-1)(n-2)(n-3)}\sum_i \left(\frac{y_i - \bar{y}}{s}\right)^4 - \frac{3(n-1)^2}{(n-2)(n-3)}$$

For variables such as weight, height or milk yield, frequency distributions are expected to be symmetric about the mean and bell-shaped. These are normal distributions. If observations follow **a normal distribution then kurtosis is equal to zero**. A distribution with positive kurtosis has a large frequency of observations close to the mean and thin tails. A distribution with a negative kurtosis has thicker tails and a lower frequency of observations close to the mean than does the normal distribution.

Excess kurtosis

The excess kurtosis is defined as kurtosis minus 3. There are 3 distinct regimes as described below:

Mesokurtic

Distributions with zero excess kurtosis are called mesokurtic, or mesokurtotic. The most prominent example of **a mesokurtic distribution is the normal distribution family**, regardless of the values of its parameters. A few other well-known distributions can be mesokurtic, depending on parameter values: for example, the binomial distribution is mesokurtic for:

$$\rho = 1/2 \pm \sqrt{1/12}$$

Leptokurtic

A distribution with positive excess kurtosis is called leptokurtic, or leptokurtotic. **"Lepto" means "slender".** In terms of shape, a leptokurtic distribution has fatter tails. Examples of leptokurtic distributions include the **Student's t-distribution**, Rayleigh distribution, Laplace distribution, exponential distribution, Poisson distribution and the logistic distribution. Such distributions are sometimes termed super-Gaussian.

Platykurtic

The coin toss is the most platykurtic distribution. **A distribution with negative excess kurtosis is called platykurtic, or platykurtotic. "Platy" means "broad".** In terms of shape, a platykurtic distribution has thinner tails. Examples of platykurtic distributions include the continuous and discrete uniform distributions, and the raised cosine distribution. The most platykurtic distribution of all is the Bernoulli distribution with p = 1/2 (for example the number of times one obtains "heads" when flipping a coin once, a coin toss), for which the excess kurtosis is −2.

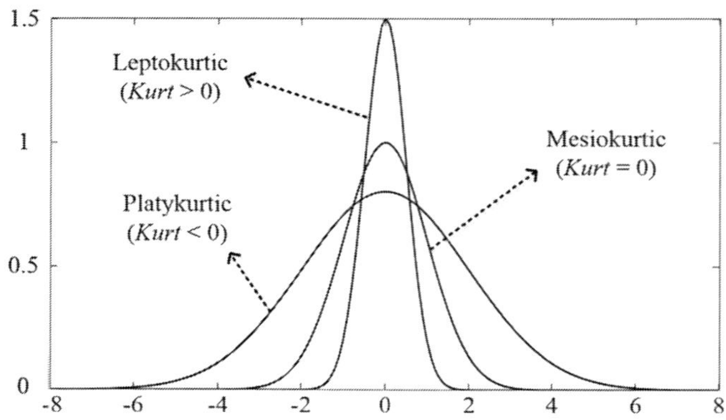

Summary

- Moments: Arithmetic mean of various powers expressed as deviation from mean in any distribution is called the moment of the distribution. Since the sum of deviations of observation from arithmetic mean is always zero, the first moment $\mu 1$ is always zero.
- The second moment about the mean μ_2 equal to variance.
- Third moment about the mean μ_3 is defined as Skewness.
- The moments can be extended to higher power in the similar way but generally the first four or up to μ_4 are defined as Kurtosis.
- For a symmetric distribution skewness is equal to zero. It is positive when the right tail is longer, and negative when the left tail is longer.
- Distributions with zero excess kurtosis are called mesokurtic, or mesokurtotic.
- A distribution with positive excess kurtosis is called leptokurtic, or leptokurtotic. "Lepto" means "slender".
- A distribution with negative excess kurtosis is called platykurtic, or platykurtotic. "Platy" means "broad".

7

Correlation and Regression

Correlation coefficient

We measure the degree of association between two variables by calculating **Pearson's product moment correlation coefficient**, usually just called the **correlation coefficient.**

It was developed by Karl Pearson from a related idea introduced by Francis Galton in the 1880s, and for which the mathematical formula was derived and published by Auguste Bravais in 1844. In 1896, Karl Pearson published his first rigorous treatment of correlation and regression in the Philosophical Transactions of the Royal Society of London.

The linear correlation coefficient (sometimes called Pearson's Correlation Coefficient), commonly denoted, is a measure of the strength of the linear relationship between two variables.

- We say that we have *perfect correlation* if all the points lie on the line; in this case, the value of the correlation coefficient takes one of its extreme values, either +1 or −1 (Figure 7a and 7b).
- We have *positive correlation* if the sign of the correlation coefficient is positive; then there is a **direct relationship** between the two variables so that as one variable increases in value, the other variable increases (Figure 7a) or there is a tendency for it to do so.
- We have *negative correlation* if the sign of the correlation coefficient is negative; then there is an **inverse relationship** between the two variables so that as one variable increases in value, the other variable decreases (Figure 7b) or there is a tendency for it to do so.
- **We have no linear association** (i.e. the **variables are *uncorrelated***) if the correlation coefficient is zero; then there is a random scatter of points with no indication of a linear relation between the variables (Figure 7c.). Note that a non-linear relationship between the variables can also give a correlation coefficient of zero.

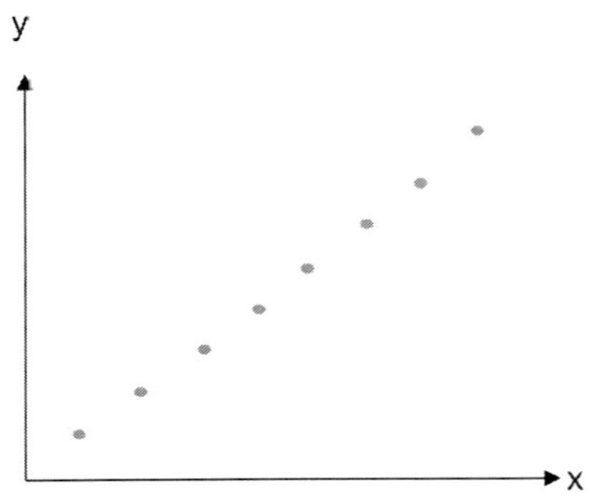

(a)

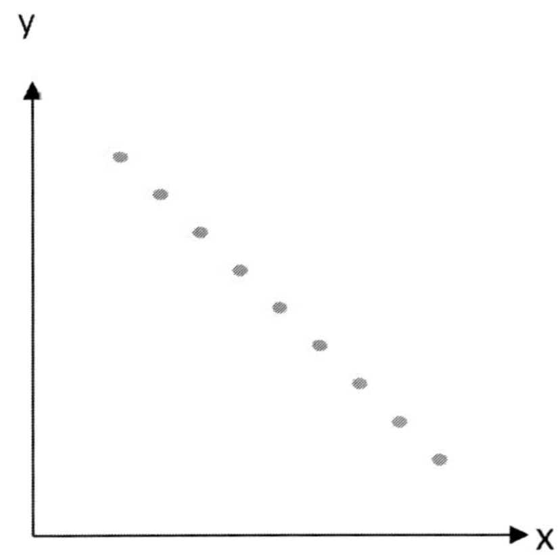

(b)

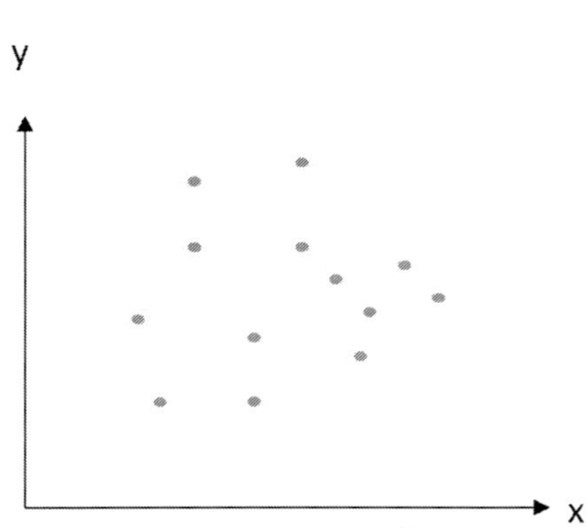

(c)

Fig. 7: Data with different correlation coefficients **(a)** perfect positive correlation coefficient r=+1 (b) perfect negative correlation coefficient **(c)** no association r=0.

If we take a random sample of n independent pairs of observations $\{(x_1, y_1), (x_2, y_2), (x_3, y_3), \ldots (x_n, y_n)\}$ on two numerical variables, x and y, then we estimate the correlation coefficient, ρ (the Greek letter rho), in the population by the sample correlation coefficient

$$\rho_{X,Y} = \frac{\operatorname{cov}(X,Y)}{\sigma_X \sigma_Y}$$

Note that

- The correlation coefficient is independent of the units of measurement of the two variables, i.e. it is dimensionless.
- The sample data provides us with the sample correlation coefficient, denoted as r.
- We can interchange x and y without affecting the value of the correlation coefficient ($r_{xy}=r_{yx}$).
- The correlation coefficient is only valid within the limits of the data in the sample.
- The absolute value of the correlation coefficient (i.e. ignoring its sign) tends to increase as the range of values of x and/or y increases, i.e. it tends to become more extreme (i.e. closer to +1 if positive and closer to −1 if negative).
- The Pearson product moment correlation coefficient provides a measure of the strength of the linear association between two variables.

Testing a hypothesis that the correlation coefficient is zero

The correlation coefficient provides a measure of the strength of the linear association between two variables. There is no linear association between the variables if the correlation coefficient is zero, and so testing the hypothesis that $\rho = 0$ is a useful exercise in correlation analysis. Note, however, that even if the **correlation coefficient is deemed significantly different from zero**, this does *not* provide evidence of a *causal* relationship between the two variables; it merely indicates that they vary together.

(a) Assumptions

There are certain assumptions that have to be satisfied if we are to test a hypothesis about the correlation coefficient, or determine the confidence interval for it. In particular:

- Both of the variables, x and y, are numerical.
- The hypothesis test that the true population correlation coefficient is zero only requires *at least one* of the two variables to be normally distributed in the population (strictly, one variable is Normally distributed with constant variance for any given value of the other variable).
- If we calculate the confidence interval for the population correlation coefficient by taking a random sample of pairs of independent observations

  $\{(x_1, y_1), (x_2, y_2), (x_3, y_3), \ldots (x_n, y_n)\}$ on two variables, x and y, then *both x and y* should be Normally distributed in the population.

(b) Approach

1. Specify the **null hypothesis** that the population correlation coefficient, ρ, is equal to zero. Generally, we adopt the alternative hypothesis that the correlation coefficient is not equal to zero.
2. **Collect** the data and **display** them in a scatter diagram from which we can discern whether a linear relationship exists between the two variables. Check the assumption that at least one (or both if a confidence interval is to be calculated) of the variables is normally distributed.
3. Calculate the sample correlation coefficient, preferably using a computer. You may find that the computer calculates a **test statistic** which has a t-distribution on $n - 2$ degrees of freedom.
4. The test statistic is

   $$t = \frac{r\sqrt{n-2}}{\sqrt{1-r^2}}$$

5. Where r is the sample correlation coefficient, and n is the number of pairs of observations in the sample. It is not difficult to calculate t-value by hand.
6. Obtain the ***P*-value**, you could refer t value to the table of the t-distribution, with $n - 2$ degrees of freedom.
7. Make a decision whether or not to reject the null hypothesis by considering the P-value. Usually, although not necessarily, we reject the null hypothesis if $P < 0.05$.
8. Calculate the confidence interval for the true correlation coefficient. Your computer package may do this automatically but, if not, we explain how

to perform the calculation manually. Although the sampling distribution of r is not Normal, the distribution of a transformed variable, z = 0.5 $\log_e\{(1 + r)/(1 - r)\}$, follows the Normal distribution, and we use this information to enable us to calculate the confidence interval for ρ.

It can be shown that the approximate 95% confidence limits for

z are $z_1 = z - 1.96/\sqrt{n\text{-}3}$

and $z_2 = z + 1.96/\sqrt{n\text{-}3}$

We then back transform, by taking exponentials, to get a confidence interval for ρ. Thus, the approximate 95% confidence interval for ρ is

$$\frac{e^{2z_1} - 1}{e^{2z_1} + 1} \text{ to } \frac{e^{2z_2} - 1}{e^{2z_2} + 1}$$

Example 1: Calculate the correlation coefficient between variable X and Y from the following data set.

Sl.No.	X	Y	X^2	Y^2	XY
1	22	18	484	324	396
2	26	20	676	400	520
3	27	23	729	529	621
4	29	24	841	576	696
5	30	22	900	484	660
6	31	21	961	441	651
7	32	25	1024	625	800
8	35	30	1225	900	1050
9	36	26	1296	676	935
10	32	27	1024	729	864
Total	$\Sigma X=300$	$\Sigma Y=236$	$\Sigma X^2=9160$	$\Sigma Y^2=5684$	$\Sigma XY=7194$

$$r_{yx} = \frac{\text{Cov.xy}}{\sqrt{\text{Var.x.Var y}}} = \frac{\Sigma XY - \Sigma X \Sigma Y / N}{\sqrt{\Sigma X^2 - \frac{(\Sigma X)^2}{N} * \Sigma y^2 - (\Sigma y)^2 / N}}$$

$$= \frac{7194 - 300x236/10}{\sqrt{(9160 - \frac{300*300}{10})\left(5684 - \frac{236*236}{10}\right)}} = \frac{7194-7080}{\sqrt{(9160-9000)(5684-5569.6)}}$$

$$\frac{114}{\sqrt{160*114.4)}} = \frac{114}{\sqrt{18304)}} = \frac{114}{135.292} = 0.8426$$

Regression coefficient

- Regression measures the average relationship between two variables.
- **Concept of regression was given by Francis Galton**

  The concept of "regression towards the mean" derived from Sir Francis Galton's work "Regression towards mediocrity in hereditary stature" publish in the year 1886 in The Journal of the Anthropological Institute of Great Britain and Ireland, 15: 246–263.
- It is denoted as 'b'.
- b_{yx} means regression of y on x, where, y is dependent variable and x independent variable.
- b_{xy} means regression of x on y, where, x is dependent variable and y is independent variable.
- In case of correlation, $r_{xy} = r_{yx}$
- In case of regression, $b_{yx} \neq b_{xy}$

Regression coefficient: It measures the amount of change in dependent variable for unit change in the independent variable.

$$b_{yx} = \frac{Cov.xy}{Var.x} = \frac{\Sigma XY - \Sigma X \Sigma Y / N}{\Sigma X^2 - (\Sigma X)^2 / N}$$

Likewise,

$$b_{xy} = \frac{Cov.xy}{Var.y} = \frac{\Sigma XY - \Sigma X \Sigma Y / N}{\Sigma Y^2 - (\Sigma Y)^2 / N}$$

Relationship between correlation and regression:

$$r_{xy} = \sqrt{bxy.byx}$$

$$b_{xy} = \frac{Cov.xy}{Var.y} \times \frac{6x}{6x} = \frac{Cov.xy}{6y.6y} \times \frac{6x}{6x}$$

$$= \frac{Cov.xy}{6y.6x} \times \frac{6x}{6y} = r.\frac{6x}{6y}$$

Similarly:

$$b_{yx} = \frac{Cov.xy}{Var.x} \times \frac{6y}{6y} = \frac{Cov.xy}{6x.6x} \times \frac{6y}{6y}$$

$$= \frac{Cov.xy}{6x.6y} \times \frac{6y}{6x} = r.\frac{6y}{6x}$$

$$b_{xy}.b_{yx} = r.\frac{6x}{6y} \times r\frac{2y}{6x} = r^2$$

$$= r^2$$

$$r = \sqrt{bxy.byx}$$

Thus, Coefficient of correlation = Geometric mean of two regression coefficients.

Properties of regression coefficient

1. Both the regression coefficients i.e., byx and bxy have the same sign. i.e. either both will be positive or both will be negative.
2. r_{xy} will also have the same sign as that of byx & bxy.
3. If one of the regression coefficients is more than one, then the other will be less than one.
4. Regression coefficient ranges from $-\infty$ to $+\infty$.
5. Regression coefficients have the **same unit as that of the unit of original observation (data).**
6. The arithmetic mean of two regression coefficients is equal or more than one [½(bxy + byx) ≥ 1].
7. The maximum value of bxy . byx = 1, because bxy.byx = r^2 & maximum value of r = 1, so r^2 = 1. Or, the arithmetic means (AM) of both regression coefficients is equal to or greater than the coefficient of correlation.

Test of regression coefficient

1. Regression coefficient is tested through t test at (n -2) df where n is the pair no. of observations.

Testing formula:

$$t_{(n\text{-}2)df} = \frac{byx}{S.E.(b)}$$

$$S.E.\ b_{yx} = \sqrt{\frac{\left[\Sigma y^2 - (\Sigma y)^2 / n\right] - b\left[\Sigma xy - \Sigma x \Sigma y / n\right]}{(n-2)\left[\Sigma x^2 - (\Sigma x)^2 / n\ \right]}}$$

Interpretation

If the calculated value of t at 5% or 1% level of significance is more than the tabulated value of t at (n-2) df and corresponding level of significance (5% or 1%) then the regression coefficient is said to be significant.

2. The F-test

For a multiple regression model with intercept, we want to test the following null hypothesis and alternative hypothesis:

H0: $\beta 1 = \beta 2 = \ldots = \beta p\text{-}1 = 0$

H1: $\beta j \neq 0$, for at least one value of j

This test is known as the overall F-test for regression.

Here are the five steps of the overall F-test for regression

1. State the null and alternative hypotheses:

   H0: $\beta 1 = \beta 2 = \ldots = \beta p\text{-}1 = 0$

   H1: $\beta j \neq 0$, for at least one value of j

2. Compute the test statistic assuming that the null hypothesis is true:

   F = MSM / MSE = (explained variance) / (unexplained variance)

3. Find a $(1 - \alpha)100\%$ confidence interval I for (DFM, DFE) degrees of freedom using an F-table or statistical software.
4. Accept the null hypothesis if $F \in I$; reject it if $F \notin I$.
5. Use statistical software to determine the p-value.

Practice Problem: For a multiple regression model with 35 observations and 9 independent variables (10 parameters), SSE = 134 and SSM = 289, test the null hypothesis that all of the regression parameters are zero at the 0.05 level.

Solution: DFE = n - p = 35 - 10 = 25 and DFM = p - 1 = 10 - 1 = 9. Here are the five steps of the test of hypothesis:

State the null and alternative hypothesis:

H0: $\beta 1 = \beta 2 = , \ldots , = \beta p\text{-}1 = 0$

H1: $\beta j \neq 0$ for some j

Compute the test statistic:

F = MSM/MSE = (SSM/DFM) / (SSE/DFE) = (289/9) / (134/25) = 32.111 / 5.360 = 5.991

Find a $(1 - 0.05)\times 100\%$ confidence interval for the test statistic. Look in the F-table at the 0.05 entry for 9 df in the numerator and 25 df in the denominator.

This entry is 2.28, so the 95% confidence interval is [0, 2.34]. This confidence interval can also be found using the R function call qf(0.95, 9, 25).

Decide whether to accept or reject the null hypothesis: 5.991 $\notin$ [0, 2.28], so reject H0.

Determine the p-value. To obtain the exact p-value, use statistical software. However, we can find a rough approximation to the p-value by examining the other entries in the F-table for (9, 25) degrees of freedom:

Level	Confidence Interval	F-value
0.100	[0, 0.900]	1.89
0.050	[0, 0.950]	2.28
0.025	[0, 0.975]	2.68
0.010	[0, 0.990]	3.22
0.001	[0, 0.999]	4.71

The F-value is 5.991, so the p-value must be less than 0.005.

Uses of Regression coefficient

1. To find out the amount of change takes place in the dependent variable for unit change in independent variables.
2. To determine the degree of resemblance between relatives.
3. To estimate heritability through regression of offspring on parent.
4. To predict the value of dependent variable for any known value of the independent variable.

Prediction equations

$$Y - \overline{Y} = b(x - \overline{x})$$

1. $y' = \overline{y} + b_{yx}(x - \overline{x})$

On the basis of correlation coefficient:

$$\mathbf{y' = \overline{} + r(x - \overline{x})}$$

$$x' = \overline{x} + r(y - \overline{y})$$

Where, y' = Predicted value of y.

x' = Predicted value of x.

$\overline{Y}$ = Average value of y variable

$\overline{x}$ = Average value of x variable

Computation of regression coefficient

On the basis of following data of x and y variables, find out regression coefficient of y on x and regression coefficient of x on y. b_{yx} and b_{xy}.

X	Y	XY	X2	Y2
10	11	no	100	121
12	12	144	144	144
14	13	182	196	169
16	14	224	256	196
18	15	270	324	225
20	16	320	400	256
22	17	374	484	289
24	18	432	576	324
26	19	449	676	361
28	20	560	784	400
30	21	630	900	441
$\Sigma X = 220$	$\Sigma y = 176$	$\Sigma XY = 3740$	$\Sigma X^2 = 4840$	$\Sigma Y^2 = 2926$

$\sum X = 220$ $\sum Y = 176$ $\sum XY = 3740$ $\sum X^2 = 4840$ $\sum Y^2 = 2926$

$(\sum X)^2 = 220*220 = 48400$ $(\sum Y)^2 = 176*176 = 30976$

$$bxy = \frac{\text{Cov.xy}}{\text{Var.}y} = \frac{\Sigma XY - \Sigma X \Sigma Y / N}{\Sigma Y^2 - (\Sigma Y)^2 / N} = \frac{3740 - 220x176/11}{2926 - 30976/11}$$

$$= \frac{3740 - 3520}{2926 - 2816} = \frac{220}{110} = 2$$

$\sum X = 220$ $\sum Y = 176$ $\sum XY = 3740$ $\sum X^2 = 4840$ $\sum Y^2 = 2926$

$(\sum X)^2 = 220*220 = 48400$ $(\sum Y)^2 = 176*176 = 30976$

$$b_{yx} = \frac{\Sigma XY - \Sigma X \Sigma Y / N}{\Sigma X^2 - (\Sigma X)^2 / N}$$

$$= \frac{3740 - 220 \text{ x } 176/11}{4840 - 48400/11}$$

$$= \frac{3740 - 3520}{4840 - 4400} = \frac{220}{440} = 1/2$$

Hence, bxy x byx = 2 x 1/2 = 1

Regression equations

$Y = a + b_X$

$\overline{y} = a + b\overline{X}$

Exercise no. 1. Find out the regression coefficient of average birth weight (y) on litter size (x) in mice from the following information. Also estimate regression coefficient of litter size on birth weight and correlation coefficient between litter size and birth weight.

Litter size (x) = 4, 5, 6, 7, 8, 9, 10, 11, 12, 13

Av. Birth wt(g) (y) = 6, 6, 5, 5, 4, 4, 3, 3, 2, 2

Answer

Sl.No.	X	Y	XY	X^2	Y^2
1	4	6	24	16	36
2	5	6	30	25	36
3	6	5	30	36	25
4	7	5	35	49	25
5	8	4	32	64	16
6	9	4	36	81	16
7	10	3	30	100	9
8	11	3	33	121	9
9	12	2	24	144	4
10	13	2	26	169	4
Total	$\sum X=85$	$\sum Y=40$	$\sum XY=300$	$\sum X^2=805$	$\sum Y^2=180$

$(\sum X)^2= 7225$ $(\sum Y)^2= 1600$

$$b_{xy} = \frac{Cov.xy}{Var.y} = \frac{\sum XY - \sum X \sum Y / N}{\sum Y^2 - (\sum Y)^2 / N} = \frac{300 - 85 * 40 / 10}{180 - 40 * 40 / 10}$$

$$= \frac{300 - 340}{180 - 160} = \frac{-40}{20} = -2$$

$$b_{yx} = \frac{\sum XY - \sum X \sum Y / N}{\sum X^2 - (\sum X)^2 / N} = \frac{300 - 85x40/10}{805 - 85*85/10} = \frac{300 - 340}{805 - 722.5} = \frac{-40}{82.5} = -0.4848$$

$r^2 = b_{xy}.\ b_{yx} = (-2)(-0.4848) = 0.9696$

$r = \sqrt{0.9696} = \pm\ 0.9847$

$$r_{yx} = \frac{Cov.xy}{\sqrt{Var.x.Var\ y}} = \frac{\sum XY - \sum X \sum Y / N}{\sqrt{\sum X^2 - \frac{(\sum X)^2}{N} * \sum y^2 - (\sum y)^2 / N}} = \frac{300 - 85 * 40 / 10}{\sqrt{20 * 82.5}}$$

$$= \frac{300 - 340}{\sqrt{1650}} = \frac{-40}{40.62} = -0.9847$$

Exercise 2. First lactation milk production of a cow was estimated to be 2800 kg. Predict the milk yield of cow in her second lactation with the following information.

i) Average 1st lactation milk yield ($\overline{X}$) = 2400 kg

ii) Average 2nd lactation milk yield ($\overline{Y}$) = 2600 kg

iii) $b_{YX} = 0.60$

$y' = \overline{y} + b_{yx}(X - \overline{X})$

$= 2600 + 0.60(2800 - 2400)$ kg

$=2600 + 0.60 \times 400$ kg

$=2600 + 240$ kg

$=2840$ kg

Summary

- The degree of association between two variables by calculating Pearson's product moment correlation coefficient, usually just called the correlation coefficient. It can take any value from −1 to +1. The value 0 means no correlation.
- Regression coefficient: It measures the amount of change in a dependent variable for a unit change in the independent variable.
- Regression coefficient ranges from $-\infty$ to $+\infty$.

Rank correlation: Spearman rank correlation coefficient as a non-parametric equivalent to the Pearson correlation coefficient used for qualitative data.

Spearman's rank correlation coefficient, also known as Spearman's ρ, was developed by Charles Spearman, a British psychologist, in 1904.

i) It measures the degree of association between the ranks of two variables.

ii) Concept given by Spearman.

iii) No unit

iv) Ranges from -1 to + 1

v) $R = 1 - \dfrac{6\sum di^2}{n(n^2 - 1)}$

Where,

$di = xi - yi$

$xi = i^{th}$ rank of x variable

yi = i^{th} rank of y variable

n = pair number of observation

$\sum$di = 0

To avoid the mathematical complexity the differences between the ranks are squared.

Example 1: Marks obtained by 10 students in Maths (X) and Statistics (Y) are given below:

X	58	64	45	51	60	62	70	55	52	56
Y	75	72	56	64	60	69	80	65	58	60

Estimation of rank correlation between the ranks of students secured for marks obtained in two subjects

Student	**Maths (Xi)**	**Rank (Xi)**	**Statistics (Yi)**	**Rank (Yi)**	**di = xi – yi**	**di^2**
1	58	5	75	2	3	9
2	64	2	72	3	-1	1
3	45	10	56	10	0	0
4	51	9	64	6	3	9
5	60	4	60	8	-4	16
6	62	3	69	4	-1	1
7	70	1	80	1	0	0
8	55	7	65	5	2	4
9	52	8	58	9	-1	1
10	56	6	61	7	-1	1
Total					$\sum$di = 0	$\sum$di2= 42

$$R = 1 - \frac{6\sum di^2}{n(n^2 - 1)}$$

$$= 1 - \frac{6x42}{10(10^2 - 1)}$$

$$= 1 - \frac{252}{10(100 - 1)}$$

$$= 1 \frac{252}{10x\ 99} - = 1 - 252/990$$

$$= 1 - 0.2545 = 0.7455$$

Coefficient of determination

The coefficient of determination or R^2 method is the proportion of the variance in the dependent variable that is predicted from the independent variable. It indicates the level of variation in the given data set.

- The coefficient of determination is the square of the correlation(r), thus it ranges from 0 to 1.
- With linear regression, the coefficient of determination is equal to the square of the correlation between the x and y variables.
- If R^2 is equal to 0, then the dependent variable cannot be predicted from the independent variable.
- If R^2 is equal to 1, then the dependent variable can be predicted from the independent variable without any error.
- If R^2 is between 0 and 1, then it indicates the extent that the dependent variable can be predictable. If R^2 of 0.15 means, it is 15 percent of the variance in the y variable is predicted from the x variable.
- The value of R^2 shows whether the model would be a good fit for the given data set. In the context of analysis, for any given per cent of the variation, it (good fit) would be different.
- For instance, in a few fields like astronomy science, R^2 is expected to be nearer to 100 %.
- But $R^2 = 0$ (minimum theoretical value), which might not be true as R^2 is always greater than 0 (by Linear Regression).

During evaluating the goodness-of-fit of simulated (Ypred) versus measured (Yobs) values, it is not appropriate to base this on the R^2 of the linear regression (i.e., Yobs= m·Ypred + b). The R^2 quantifies the degree of any linear correlation between Yobs and Ypred, while for the goodness-of-fit evaluation only one specific linear correlation should be taken into consideration: Yobs = 1·Ypred + 0 (i.e., the 1:1 line).

The most general definition of the coefficient of determination is

$$R^2 = 1 - \frac{SS_{res}}{SS_{tot}}$$

Where,

$$SS_{res} + SS_{reg} = SS_{tot}$$

$$R^2 = \frac{SS_{reg}}{SS_{tot}} = \frac{SS_{reg}/n}{SS_{tot}/n}$$

In the best case, the model values exactly match the observed values, which results in SSres=0, R^2=1

A baseline model, which always predicts y will have R^2=0.

Properties of coefficient of determination

- It helps to get the ratio of how a variable which can be predicted from the other one, varies.
- If we want to check how clear it is to make predictions from the data given, we can determine the same by this measurement.
- It helps to find Explained variation / Total Variation
- It also lets us know the strength of the association(linear) between the variables.
- If the value of r^2 gets close to 1, The values of y become close to the regression line and similarly if it goes close to 0, the values get away from the regression line.

• It helps in determining the strength of association between different variables.

8

Probability

A. Definitions of probability

1. A probability relies on having an understanding of the theoretical **model** defining the set of all possible outcomes of a trial; we evaluate the probability solely on the basis of this model, without recourse to performing the experiment at all. It is often called an **a priori** probability. So, for example, we know that there are two equally likely outcomes when an unbiased coin is tossed: either a head or a tail. This is the model from which we can deduce that the probability of a defined event, obtaining a head, say, is 1/2 = 0.5.
2. The next approach to defining a probability, and the one commonly used in statistical inference, is to regard a probability as the proportion of times a particular outcome (the event) will occur in a very large number of 'trials' or 'experiments' performed under similar conditions.

The result of any one trial should be independent of the result of any other trial, so whether or not the event occurs in any one trial should not affect whether or not the event occurs in any other trial.

Properties of a probability

1. It is clear that, since a probability can be defined as a relative frequency or a proportion, its numerical value must be equal to or lie between 0 and 1.
   - A probability of 0 means that the event *cannot* occur.
   - A probability of 1 (unity) means that the event *must* occur.
2. We often convert probabilities into percentages (with a range 0–100%) or express them as ratios (e.g. a one-in-three chance of an event occurring).
3. Sometimes we focus our interest not on a particular event occurring but on that event not occurring, i.e. on the **complementary event**. It follows from the properties of a probability that the probability of the event not

occurring is **1 minus the probability of the event occurring**. E.g. So, if the probability of a kitten affected feline viral rhinotracheitis after vaccination at 9 and 13 weeks of age is 0.04 (in a particular location and time), then the probability of being adequately protected is 1 - 0.04=0.96.

Rules of probability

For simplicity, we define for only two events, A and B, but they can be extended to multiple events.

1. **Addition rule:** When two events are mutually exclusive, implying that the two events cannot occur at the same time, then the probability of either of the two events occurring is the sum of the probability of each event.

Thus, Pr(A or B) = Pr(A) + Pr(B)

For example, assuming that we have a carton of 50 dog biscuits, with 10 of each of five different shapes, the probability of picking either a diamond shape or a round shape from the carton is the sum of the probability of a diamond (10/50 = 1/5) and the probability of a round (10/50 = 1/5) which is 1/5+1/5=2/5 or 0.4.

But if events are not mutually exclusive then rule of addition is modified as

Pr(A or B) = Pr(A) + Pr(B) – P (AB)

Example: Probability of drawing a king or a club from a pack of cards will be:

P (king or club) = 4/52 +13/52 – (4/52) (13/52) = 4/13

2. **Multiplication rule**: When two events are independent, so that the occurrence or non occurrence of one event does not affect the occurrence or non-occurrence of the other event, then the probability of both events occurring is the product of the individual probabilities.

Thus,

Pr(A or B) = Pr(A) × Pr(B)

Example 1, If we have two cartons of dog biscuits, as in the addition rule example, the probability of picking a diamond shape from both cartons is 1/5 × 1/5, equal to 1/25 or 0.04.

Example 2: A bag contains 5 white and 3 black balls. Find the probability that 2 balls are successively drawn is white and not replaced.

P (first white) = 5/8

P (second white if first one is not replaced) = 4/7 P (WW) = (5/8) (4/7) = 5/14

When **two events are *not independent***, we have to adopt a different rule, which relies on an understanding of **conditional probability**. **The probability of an event B occurring when we know that A has already occurred is called the conditional probability of B, and is written as *Pr*(B given A)** or ***Pr*(B|A).** Thus event B is dependent on A. If we have two such dependent events, the probability of both events occurring is equal to the probability of one of them occurring times the conditional probability of the other occurring. So,

***Pr*(A or B) = *Pr*(A) × *Pr*(B given A)**

For example, if we have a carton of dog biscuits as in the addition rule example, the probability of picking a second diamond shape after we have already picked one diamond shape (and given it to Max to eat!) is equal to the probability of picking the first diamond shape (10/50 = 1/5) times the probability of picking the second diamond shape out of the remaining 49 biscuits (9/49), i.e. (1/5)*(9/49) or it is 0.037.

3. Permutation (arrangement of object)

The permutation of n different objects taken r at a time is denoted by

$^nPr = n!/(n-r)!$

Example:

A,B,C take two at a time

If n=3, r =2, $^3P2 = 3!/(3-2)! = (3*2*1)/1 = 6$

AB, BC, CA, BA,CB AC

4. Combination (without group selection)

The combination of n different objects taken r at a time is denoted by

$^nCr = n!/r!(n-r)! = {}^nPr/r!$

Example:

Total number of combinations of 3 letters taking 2 at a time.

$^3C2 = 3!/2!(3-2)! = (3*2*1)/(2*1)(1) = 6/2 = 3$

5. **Compound events (Probability)**

A compound event is an event composed of two or more events. Consider two events *A* and *B*. The compound event such that **both events *A* and *B* occur is called the *intersection of the events,* denoted by $A \cap B$.** The compound event such that either event *A* or event *B* occurs is called the *union of events,* denoted by $A \cup B$. The probability of an intersection is $P(A \cap B)$ and the probability of union is $P(A \cup B)$. Also:

$$P(A \cup B) = P(A) + P(B) - P(A \cap B)$$

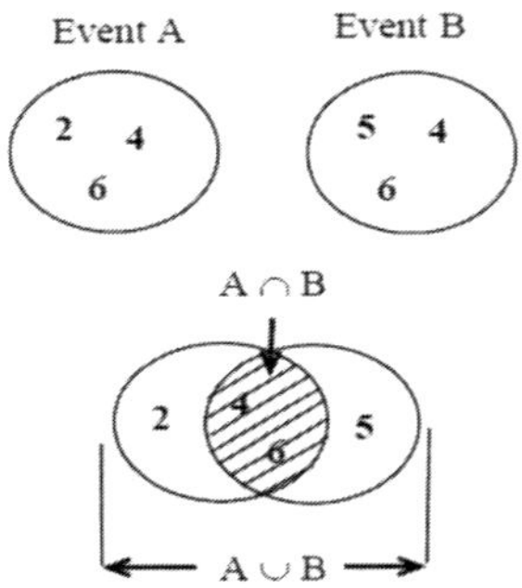

Fig. Intersection and union of two events, events *A* and *B*.

6. **Conditional probability,** a conditional probability is a probability that **an event *B* will occur if it is known that an event *A* has already occurred.** The conditional probability of B given A is calculated by using the formula:

$$P(B / A) = \frac{P(A \cap B)}{P(A)}$$

and consequently the probability that both events occur is:

$$P(A \cap B) = P(A)\, P(B|A)$$

If events *A* and *B* are independent then:

$$P(B \mid A) = P(B) \text{ and } P(A \mid B) = P(B)$$

$$P(A \cap B) = P(A)\, P(B)$$

An example of dependent events: From a deck of 52 playing cards we draw two cards. What is the probability that both cards drawn are aces?

The first draw is event *A* and the second is event *B*. Recall that in a deck there are four aces. The probability that both are aces is $P(A \cap B)$. The events are obviously dependent, namely drawing of the second card depends on which card has been drawn first.

$P(A = \text{Ace}) = (4/52) = (1/13)$.

$P(B = \text{Ace} \mid A = \text{Ace}) = (3/51)$, that is, if the first card was an ace, only 51 cards were left and only 3 aces.

Thus,

$P(A \cap B) = P(A)\, P(B|A) = (4/52)\,(3/51) = (1/221)$.

The probability of drawing two aces is (1/221).

7. Bayes' theorem

Bayes theorem is useful for stating the probability of some event A, if there is information about the probability of some event E that happened after the event A.

Consider two cages $K1$ and $K2$, in the first cage there are three mice, two brown and one white, and in the second there are two brown and two white mice. Each brown mouse is designated with the letter B, and each white mouse with the letter W.

Cage $K1$	Cage $K2$
B,B,W	B,B,W,W

$$P(K_1 / B) = P(K_1 \cap B) = \frac{P(K_1)P(B / K_1)}{P(K_1)P(B / K)_1 + P(K_2)P(B / K_2)}$$

Generally, there is an event A with k possible outcomes $A1, A2,..., Ak$, that are independent and the sum of their probabilities is 1, ($\Sigma i\ P(Ai) = 1$). Also, there is an event E, that occurs after event A. Then:

$$P(Ai / E) = \frac{P(A_i \cap E)}{P(E)} = \frac{P(A_i)P(E / A_i)}{P(A_1)P(E / A_1) + P(A_2)P(E / A_2) + ... + P(A_k)P(E / A_k)}$$

Problem: For artificial insemination of some large dairy herd semen from two bulls are utilized. Bull 1 has been used on 60% of the cows, and bull 2 on 40%. We know that the percentage of successful inseminations for bull 1 and bull 2 are 65% and 82%, respectively. For a certain calf the information about its father has been lost. What is the probability that the father of that calf is bull 2?

We define:

$P(A1) = 0.6$ is the probability of having used bull 1

$P(A2) = 0.4$ is the probability of having used bull 2

E = the event that a calf is born (because of successful insemination)

$P(E \mid A1) = 0.65$ = the probability of successful insemination if bull 1

$P(E \mid A2) = 0.82$ = the probability of successful insemination if bull 2

$$P(A_2 / E) = \frac{P(A_2 \cap E)}{P(E)} = \frac{P(A_2)P(E / A_2)}{P(A_1)P(E / A_1) + P(A_2)P(E / A_2)} = \frac{(.4)(.82)}{(.6)(.65) + (.4)(.82)} = 0.457$$

Thus, the probability that the father of that calf is bull 2 is 0.457

9

Probability Distributions Binomial, Poisson, Chi-Square and Normal

A probability distribution is a statistical function that describes all the possible values and likelihoods that a random variable can take within a given range. This range will be restricted between the minimum and maximum possible values.

The probability distribution shows how the set of all possible mutually exclusive events are distributed, and can be presented as an *equation*, a *chart* or a *table*.

A variable which can take different values with given probabilities is called a **random variable**.

There are numerous probability distributions which may be distinguished by whether the random variable is **discrete**, taking only a finite set of possible values, or **continuous**, taking an infinite set of possible values in a range of values.

A **discrete random variable** with only two possible values is called a **binary variable**, e.g. pregnant or not pregnant, diseased or healthy.

Probability distributions for discrete random variables

The probability distribution for a discrete random variable y is the table, graph or formula that assigns the probability $P(y)$ for each possible value of the variable y. The probability distribution $P(y)$ must satisfy the following two assumptions:

1) $0 \leq P(y) \leq 1$

   The probability of each value must be between 0 and 1, inclusively.

2) $\Sigma(all\ y)\ P(y) = 1$

The sum of probabilities of all possible values of a variable y is equal to 1.

1. Binomial distribution

Assume a single trial that can take only two outcomes, for example, Yes and No, success and failure, or 1 and 0. Such a variable is called a binary or Bernoulli variable.

Now assume that such a single trial is repeated n times. A binomial variable y is the number of successes in those n trials. It is the sum of n binary variables. The binomial probability distribution describes the distribution of different values of the variable y $\{0, 1, 2, \ldots, n\}$ in a total of n trials. Characteristics of a binomial experiment are:

1) The experiment consists of n equivalent trials, independent of each other
2) There are only two possible outcomes of a single trial, denoted with Y (yes) and N (no) or equivalently 1 and 0
3) The probability of obtaining Y is the same from trial to trial, denoted with p. The probability of N is denoted with q, so $p + q = 1$
4) The random variable y is the number of successes (Y) in the total of n trials.

The probability distribution of a random variable y is determined by the parameter p and the number of trials n:

$$P(y) = \binom{n}{y} p^y q^{n-y}$$

y = 0, 1, 2,……n

Where,

P = the probability of success in a single trial

q= 1-p =the probability of failure in a single trial

The expectation and variance of a binomial variable are:

$E(y) = \mu = np$ and $Var(y) = \sigma^2 = npq$

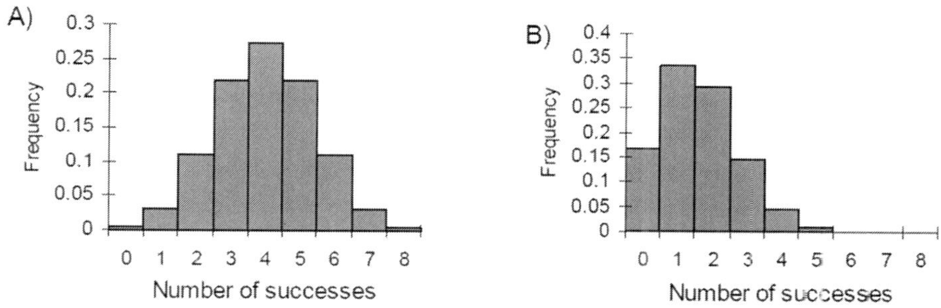

Fig. 9.1: The shape of the distribution depends on the parameter p. The binomial distribution is symmetric only when $p = 0.5$, and asymmetric in all other cases. The figures A and B, present two binomial distributions for $p = 0.5$ and $p = 0.2$ with n = 8.

Binomial distribution can be elaborated as follows

$P(x) = {}^{n}C_{x} . p^{x} (1 - p)^{n-x}$

Where,

- n = Total number of events
- r (or) x = Total number of successful events.
- p = Probability of success on a single trial.
- ${}^{n}C_{r} = [n!/r!(n-r)!]$
- 1 – p = Probability of failure.

Example: A coin is tossed 12 times. What is the probability of getting exactly 7 heads?

Solution

Given that a coin is tossed 12 times. (i.e) n= 12

Thus, a probability of getting head in single toss = ½. (i.e) p = ½.

So, 1-p = 1-½ = ½.

We know that the binomial probability distribution is $P(r) = nCr \cdot p^{r} (1 - p)^{n-r}$.

Now, we have to find the probability of getting exactly 7 heads.(i.e) r = 7.

Substituting the values in the binomial distribution formula, we get

$P(7) = 12C7 \cdot (½)^{7} (½)^{12-7}$

$P(7) = 792 \cdot (½)^{7} (½)^{5}$

P(7) = 792.(½)^{12}

P(7) = 792 (1/4096)

P(7) = 0.193

Therefore, the probability of getting exactly 7 heads is 0.193.

2. Poisson distribution

The Poisson distribution is a model for the relative frequency of rare events and data defined as counts and often is used for determination of the probability that some event will happen in a specific time, volume or area.

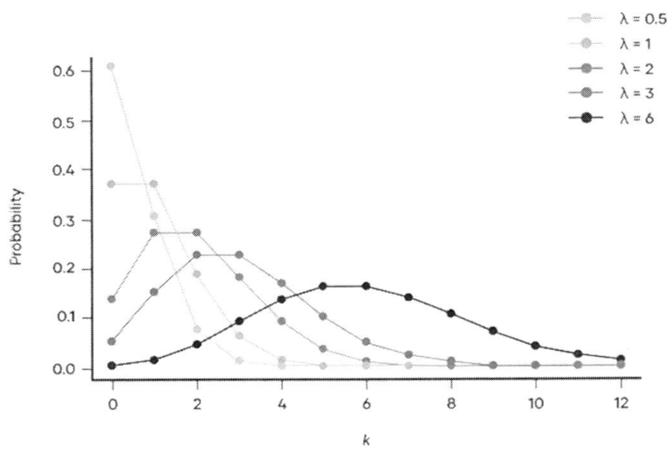

Fig. 9.2: A Poisson distribution graph with different **values** of **lambda**. The horizontal axis is the index k, the number of occurrences. λ is the expected rate of occurrences. The vertical axis is the probability of k occurrences given λ.

For example, the number of microorganisms within a microscope field, or the number of mutations or distribution of animals from some plot may have a Poisson distribution. A Poisson random variable y is defined as how many times some event occurs in a specific time, or given volume or area. If we know that each single event occurs with the same probability, that is, the probability that some event will occur is equal for any part of time, volume or area, and the expected number of events is λ, then the probability function is defined as:

$$P(y) = \frac{e^{-\lambda}\lambda^{y}}{y!}$$

where λ is the average number of successes in a given time, volume or area, and e is the base of the natural logarithm (e = 2.71828). Often, instead of the expected number, the proportion of successes is known, which is an estimate of the probability of success in a single trial (p). When p is small and the total number of trials (n) large, the binomial distribution can be approximated with a Poisson distribution, $\lambda = np$. **A characteristic of the Poisson variable is that both the expectation and variance are equal to the parameter λ:**

$E(y)=\mu=\lambda,\ Var(y)=\sigma^2=\lambda$

Example 1: In a population of mice 2% have cancer. In a sample of 100 mice, what is the probability that more than one mouse has cancer?

Answer

$\mu=\lambda=100(0.02)=2$

$P(y)=e^{-2}2^y/y!$

$P(y>1)=1-P(y=0)-P(y=1)=1-0.1353-0.2706=0.5941$

The probability that in the sample of 100 mice more than one mouse has cancer is 0.5941.

Example 2

An average of 0.61 soldiers died by horse kicks per year in each Prussian army corps. You want to calculate the probability that exactly two soldiers died in the VII Army Corps in 1898, assuming that the number of horse kick deaths per year follows a Poisson distribution.

Calculation

The specific army corps (VII Army Corps) and year (1898) don't matter because the probability is constant.

$k = 2$ deaths by horse kick

$\lambda = 0.61$ deaths by horse kick per year

$e = 2.718$

$$P(X=k)=\frac{e^{-\lambda}\lambda^k}{k!}$$

$$P(X=2)=\frac{(2.718^{-0.61})(0.61^2)}{2!}$$

$$P(X=2)=\frac{(0.54339)(0.3721)}{2}$$

P (X = 2) = 0.101

The probability that exactly two soldiers died in the VII Army Corps in 1898 is 0.101.

3. *Chi*-square distribution (Test for goodness of fit)

It is used to describe the distribution of a sum of squared random variables. It is also used to test the goodness of fit of a distribution of data, whether data series are independent, and for estimating confidences surrounding variance and standard deviation for a random variable from a normal distribution.

Imagine taking a random sample of a standard normal distribution (Z). If you squared all the values in the sample, you would have the chi-square distribution with k = 1.

$\chi^2 1 = (Z)^2$

Consider a set of standard normal random variables zi, (i = 1,…, v), that are identical and independently distributed with the mean μ = 0 and standard deviation σ = 1. Define a random variable:

$\chi^2 = \Sigma i\, z^2 i,\ i = 1,\ldots, v$

The variable χ^2 has a *chi*-square distribution with v degrees of freedom. The shape of the *chi*-square distribution depends on degrees of freedom. Figure 9.3 shows *chi*-square

density functions with 2, 6 and 10 degrees of freedom.

The expectation and variance of a χ^2 variable are:

$E\,[\chi^2] = v$ and $Var\,[\chi^2] = 2v$

$$\chi^2=\sum\frac{(O_i-E_i)^2}{E_i}$$

χ^2 = chi squared

Oi = observed value

Ei = expected value

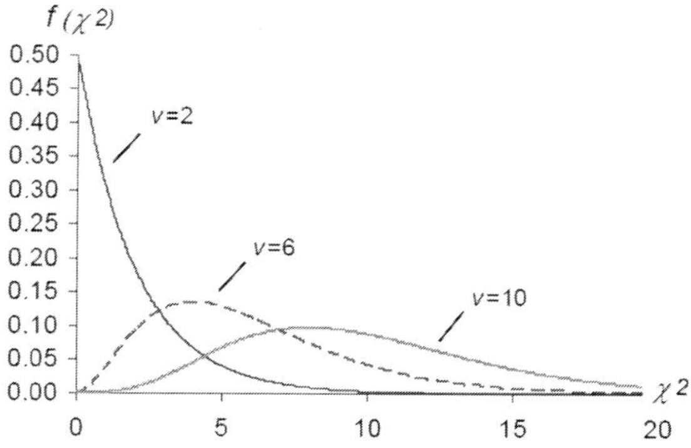

Fig. 9.3: The density functions of χ^2 variables with v=2, v=6 and v=10 **degrees** of freedom.

Where, k is greater than two, the chi-square distribution is hump-shaped. The curve starts out low, increases, and then decreases again. There is low probability that χ^2 is very close to or very far from zero. The most probable value of χ^2 is $\chi^2 - 2$. When k is only a bit greater than two, the distribution is much longer on the right side of its peak than its left (i.e., it is strongly right-skewed).

As k increases, the distribution looks more and more similar to a normal distribution. In fact, when k is 90 or greater, a normal distribution is a good approximation of the chi-square distribution.

The χ^2 test is applied in case of enumeration statistics only. Enumeration statistics are obtained by counting the individuals for descriptive characteristics, which are discrete in nature, like color of flower (yellow or pink), inoculation (number of individuals inoculated or not inoculated). It is a non-parametric test (no population parameter is required).

The χ^2 test required fulfillment of following conditions:

a) Number of observations should be sufficiently & reasonably large (at least 50) for normal distribution of difference between observed and expected frequencies.

b) **Number theoretical cell frequency should be very small (at least 5).**

c) Constraints should be linear i.e. should not involve power of frequencies.

If frequency of any cell is less than 5, the two adjacent classes are combined, if frequency still remains less than 5, three classes are combined.

If calculated 'χ^2' value is less than table 'χ^2' value at k-1 degrees of freedom then fit is good.

Example: In an experiment on pea breeding, Mendel observed the following frequencies of seed:

Round and yellow 320

Wrinkled and yellow 104

Round and green 110

Wrinkled and green 31

Test whether the observed frequencies are according the typical dihybrid ratio (9:3:3:1)?

Solution

1. H0 : Observed frequency = Expected frequency

   HA : Observed frequency $\neq$ Expected frequency

2. Assumption: Sample should be drawn randomly

   Population should be normally distributed.

3. The **tabulated value** of 'χ^2' at 5 % level of significance and 3 degrees of freedom is 7.815.

4. Calculation:

Classes	O	E	O-E	$(O-E)^2$	$(O-E)^2/E$
Round and yellow	320	317.8	2.2	4.84	0.015
Wrinkled and yellow	104	105.9	1.9	3.61	0.034
Round and green	110	105.9	4.1	16.81	0.16
Wrinkled and green	31	35.4	4.4	19.36	0.54

'χ^2' = Σ $(O-E)^2/E$ = 0.749

5. **Conclusion:** The calculated value of 'χ^2' is less than table 'χ^2' value that indicates the correspondence between observed and expected frequency. or H0 is accepted.

Testing the independence of attributes

When the data are classified in two ways (dichotomy) then we use this test for testing the independence of attributes. Such kind of data are arranged in the form of m x n contingency table as follows:

a	b	$R_1 = a + b$
c	d	$R_2 = c + d$
$C_1 = a + c$	$C_2 = b + d$	$N = C_1+C_2 = R_1+R_2$

The test statistic 'χ^2' is calculated as follows: $\chi^2 = (ad- bc)^2 N/R1R2C1C2$

Degree of freedom = (m-1) (n-1)

m = Number of rows

n = Number of columns

- If calculated 'χ^2' value is less than table 'χ^2' value at (m-1) (n-1) degrees of freedom then null hypothesis (H_0) of independence of attributes is accepted.
- If any of the cell frequency is less than 5, Yates's correction for continuity may be used to sharpen the approximation. this is accomplished by using the following formula:

$\chi^2 = [(ad\text{-}bc)\text{-}N/2]^2 N/R_1R_2C_1C_2$

It should be noted that this correction should not be used when m>2 and n>2.

Example: In an experiment with immunization of cattle for tuberculosis, the following results were obtained:

Classes	Affected	Unaffected
Inoculated	9	416
Not inoculated	14	291

Examine the effect of vaccine in controlling susceptibility to tuberculosis.

Solution

Expected value of any cell is calculated as: $f_1 = R_i \times C_i / N$

Where, R_i and C_i are the total number of observations in row and column of that particular cell. The 'χ^2' is calculated with the help of following formula:

'χ^2' = $\Sigma (O\text{-}E)^2/E$

By short cut method chi-square value is calculated as follows:

$\chi^2 = (ad\text{-}bc)^2N/R_1R_2C_1C_2$

$= (9 \times 291 - 416 \times 14)^2 \times 730 / 425 \times 305 \times 23 \times 707$

$= 3.55$

Table value at 1 degree of freedom is 3.84 that is more than calculated value.

So, we accept the null hypothesis and conclude the vaccine is not effective.

4. Normal (or Gaussian) distribution

Abraham De Moivre (1733), Gauss in 1809 and by Laplace in 1812 described normal distribution. The **Normal** or **Gaussian distribution**, named after C. F. Gauss, an 18th century German mathematician.

Properties of the normal distribution

I. The normal distribution is completely described by two parameters: the mean and the standard deviation. These are usually denoted by the Greek letters μ and σ, respectively.

II. It is unimodal and continuous variable.

III. It is symmetrical about its mean. This implies that the curve to the right of the mean is a mirror image of the curve to the left of the mean. It is often described as 'bell-shaped'.

IV. Its mean, median and mode are all equal.

V. standard normal variable

$z = (x-\mu)/\sigma$

Normal (or Gaussian) distribution

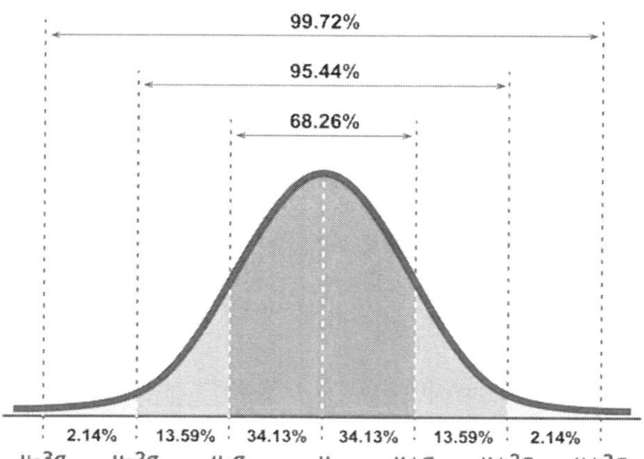

- If the standard deviation remains unchanged, increasing the value of the mean shifts the curve horizontally to the right. Conversely, decreasing the value of the mean shifts the curve horizontally to the left.

- A decrease in the standard deviation of the curve makes the curve thinner, taller and more peaked. Conversely, an increase in the standard deviation makes the curve fatter, shorter and flatter.

Standardized normal deviate (SND)

- Convert this Normal distribution into a Standard Normal distribution which has a mean of 0 and a standard deviation of 1 (unity). This is the distribution of a new variable, z, which is called a Standardized Normal Deviate (SND).
- In general terms $z = (X - \mu) / \sigma$
- Any point (x) from a normal distribution can be converted to the standard normal distribution (z) with the formula z = (x-mean) / standard deviation.
- z for any particular x value shows how many standard deviations x is away from the mean for all x values.
- Where, X is a normal random variable, μ is the mean of X, and σ is the standard deviation of X.
- The standard normal distribution is a normal distribution with mean $\mu = 0$ and standard deviation $\sigma = 1$. The letter Z is often used to denote a random variable that follows this standard normal distribution.
- A z-score is measured in units of the standard deviation.

  The mean for the standard normal distribution is zero, and the standard deviation is one. The transformation $z=(x-\mu)/\sigma$ produces the distribution $Z \sim N(0, 1)$.
- The limits $(\mu - \sigma)$ and $(\mu + \sigma)$ contain 68.3% of the distribution.
- The limits $(\mu - 1.96\sigma)$ and $(\mu + 1.96\sigma)$ contain 95% of the distribution. This fact is often used in the calculation of a reference range.
- The limits $(\mu - 2.58\sigma)$ and $(\mu + 2.58\sigma)$ contain 99% of the distribution.

The density function of normal distribution is

The probability density function of normal or gaussian distribution is given by,

$$f(x,\mu,\sigma) = \frac{1}{\sigma\sqrt{2\pi}} e^{\frac{-(x-\mu)^2}{2\sigma^2}}$$

Where,

x is the variable

μ is the mean

σ is the standard deviation

The normal distribution doesn't even bother about the range. The range can also extend to $-\infty$ to $+\infty$ and still we can find a smooth curve.

How to use the Z Table

A z-table is a table that tells you what percentage of values fall below a certain z-score in a standard normal distribution.

A z-score simply tells you how many **standard deviations** away an individual data value falls from the mean. It is calculated as:

z-score = $(x - \mu) / \sigma$

Where:

- x: individual data value
- μ: population mean
- σ: population standard deviation

This tutorial shows several examples of how to use the z table.

Example 1

The scores on a certain college entrance exam are normally distributed with mean $\mu = 82$ and standard deviation $\sigma = 8$. Approximately what percentage of students score less than 84 on the exam?

Step 1: Find the z-score.

First, we will find the z-score associated with an exam score of 84:

z-score = $(x - \mu) / \sigma = (84 - 82) / 8 = 2 / 8 = 0.25$

Step 2: Use the z-table to find the percentage that corresponds to the z-score.

Next, we will look up the value 0.25 in the z-table:

z	0	0.01	0.02	0.03	0.04	0.05	0.06	0.07	0.08	0.09
0.0	0.5000	0.5040	0.5080	0.5120	0.5160	0.5199	0.5239	0.5279	0.5319	0.5359
0.1	0.5398	0.5438	0.5478	0.5517	0.5557	0.5596	0.5636	0.5675	0.5714	0.5753
0.2	0.5793	0.5832	0.5871	0.5910	0.5948	0.5987	0.6026	0.6064	0.6103	0.6141
0.3	0.6179	0.6217	0.6255	0.6293	0.6331	0.6368	0.6406	0.6443	0.6480	0.6517
0.4	0.6554	0.6591	0.6628	0.6664	0.6700	0.6736	0.6772	0.6808	0.6844	0.6879
0.5	0.6915	0.6950	0.6985	0.7019	0.7054	0.7088	0.7123	0.7157	0.7190	0.7224
0.6	0.7257	0.7291	0.7324	0.7357	0.7389	0.7422	0.7454	0.7486	0.7517	0.7549
0.7	0.7580	0.7611	0.7642	0.7673	0.7704	0.7734	0.7764	0.7794	0.7823	0.7852
0.8	0.7881	0.7910	0.7939	0.7967	0.7995	0.8023	0.8051	0.8078	0.8106	0.8133
0.9	0.8159	0.8186	0.8212	0.8238	0.8264	0.8289	0.8315	0.8340	0.8365	0.8389
1.0	0.8413	0.8438	0.8461	0.8485	0.8508	0.8531	0.8554	0.8577	0.8599	0.8621
1.1	0.8643	0.8665	0.8686	0.8708	0.8729	0.8749	0.8770	0.8790	0.8810	0.8830
1.2	0.8849	0.8869	0.8888	0.8907	0.8925	0.8944	0.8962	0.8980	0.8997	0.9015
1.3	0.9032	0.9049	0.9066	0.9082	0.9099	0.9115	0.9131	0.9147	0.9162	0.9177
1.4	0.9192	0.9207	0.9222	0.9236	0.9251	0.9265	0.9279	0.9292	0.9306	0.9319
1.5	0.9332	0.9345	0.9357	0.9370	0.9382	0.9394	0.9406	0.9418	0.9429	0.9441
1.6	0.9452	0.9463	0.9474	0.9484	0.9495	0.9505	0.9515	0.9525	0.9535	0.9545

Conclusion

So approximately 59.87% of students score less than 84 on this exam.

Important takeaways from chapter

- Probability: Probability is a measure of the likelihood of an event occurring. It is typically expressed as a number between 0 and 1, where 0 represents an impossible event, and 1 represents a certain event. For example, the probability of getting heads when flipping a fair coin is 0.5.
- Random variables: A random variable is a variable that can take on different values with certain probabilities. It is used to model uncertain or random phenomena. For example, in a dice roll, the number that comes up is a random variable.
- Trials- Repetition of experiment
- Single trial- To perform the experiment once
- Sample space- The set of all possible outcomes of a probability experiment
- Sample point- Elements of sample space {Elements of sample space are outcomes)
- Outcome- Result of a single trial of a probability experiment
- Event- A collection of outcomes for the experiment, any subset of the sample space.

- Simple event- We consider the probability of a single event happening. In compound events we consider joint occurrence of two or more events.
- Equally likely events- Two or more events are equally likely events if both events have the same chance of occurrence.
- Mutually exclusive events- Two or more events are mutually exclusive events if no two of them have outcomes in common (those that cannot occur simultaneously).
- Independent events: Event B is said to be independent of event A if P (BIA) = P (B) {the occurrence of some does not affect the probabilities of the others occurring).
- Exhaustive Events: Events are said to be exhaustive when their totality includes all the possible outcomes of a random experiment. (Dice-possible outcomes 1,2, 3...6 and hence the exhaustive number of cases is 6).
- The addition law of probability (or sum rule): This law deals with the probability of the union of two or more events. It states that the probability of the union of two events A and B is given by $P(A \cup B) = P(A) + P(B) - P(A \cap B)$, where $P(A \cap B)$ is the probability of the intersection of A and B.
- The multiplication law of probability (or product rule): This law deals with the probability of the joint occurrence of two or more events. For independent events A and B, it states that the probability of both A and B happening is given by $P(A \cap B) = P(A) * P(B)$.
- Complement rule: This rule states that the probability of the complement of an event A (the event not A) is given by $P(A') = 1 - P(A)$.
- Conditional Probability: The probability of an event happening given that another event has occurred.
- Bayes' theorem: Bayes' theorem is a fundamental law of conditional probability that allows you to update your beliefs about an event based on new evidence. It is particularly important in Bayesian statistics and inference. The theorem is given by $P(A \mid B) = [P(B \mid A) * P(A)] / P(B)$, where $P(A \mid B)$ is the probability of event A occurring given that event B has occurred.
- A probability distribution consists of the values of a random variable and their corresponding probabilities. It can be of two types based on variable associated i.e. discrete and continuous

- Discrete probability distribution- Binomial, poissons, Geometric, Multinomial
- Continuous probability distribution- Normal, Student's-t, Chi Square, F
- Binomial distribution- By James Bernoulli in 1700 A.D.
    - In a random experiment there is two possible results each are repeated in time, such trials (n) are independent to each other, and probability of success and failure (p & q) remain as same in each trial then the probability of getting (x) success can be calculated by following formula $\mathbf{P(x) = {}^{n}C_{x} . p^{x} (1 - p)^{n-x}}$
    - **Properties of binomial distribution**
    - Parameters: n and p
    - $X \sim B(n, p)$
    - Mean = np and variance = npq.
    - Mean is always greater than variance.
    - Additive property
- **Poisson distribution- Simen Denis Poisson (1837) A.D.**
    - It is used for rare events.
    - It is a limiting form of the binomial distribution. In which n, becomes very large then binomial distribution is converted into Poisson distribution.
    - Probability of getting Success (r) can be obtained by using following formula

      $$P(y) = \frac{e^{-\lambda} \lambda^{y}}{y!}$$

      which is called as Poisson distribution.
    - **Properties of binomial distribution**
    - One parameter = λ
    - $X \sim P(\lambda)$.
    - Mean = variance = λ
    - Additive property

- **Normal distribution- Abraham De Moivre (1733), Gauss (1809) and Laplace (1812)**
  - A continuous random variable X is said to have a normal distribution with parameters μ (mean) and σ^2 (variance) if its probability density function is given by:

    $$f(x) = \frac{1}{\sigma\sqrt{2\pi}} \exp\left(-\frac{(x-u)^2}{2\sigma^2}\right)$$
  - Properties of binomial distribution-
  - The curve is bell shaped curve with two Parameters, μ and σ.
  - The distribution curve is symmetrical about the mean μ and falls rapidly on both sides tailing off asymmetrically to the X-axis in both the directions (i.e., the X – axis is tangent to the curve at infinity).
  - Mean, median and mode of the normal distribution coincide.
  - In a normal distribution, β1= 0 and β2= 3.
  - Linear combination of normal variates is also distributed normally.
  - The point of inflection occurs at $X = \mu \pm \sigma$

In a normal distribution, the values of β1 and β2 are as follows:

β1: The skewness (or symmetry) of a normal distribution is equal to zero (0).

β2: The kurtosis of a normal distribution is equal to three (3). Kurtosis measures how peaked or flat the distribution is. A value of 3 indicates a normal distribution. In general, to "normalize" the kurtosis statistic, 3 is subtracted from the statistic before reporting it

10

Hypothesis Testing

We can categorize statistical theory into two general parts.

1. Firstly, there is **descriptive statistics** which uses the appropriate tools, typically tables, diagrams and/or numerical measures, to describe a data set and provide a summary of its distribution.
2. In addition, there is **inferential statistics** which is concerned with drawing conclusions about a population using information obtained from a representative sample selected from it.

   (a) One aspect of statistical inference is the **estimation of a population parameter** by the appropriate sample statistic (e.g. the population mean by the sample mean).

   (b) The second aspect of inferential statistics is **hypothesis testing**. In this case, we examine a hypothesis, framed in terms of the parameters in one or more populations.

Estimation is concerned with ***description*** whereas hypothesis testing is ultimately concerned with ***decision***.

Basic concepts of hypothesis testing

Hypothesis testing is a process that is concerned with **making conclusions** about the population using the information obtained from its sample.

Decision about the population is correct?

One randomly selected sample from the population is unlikely to be exactly the same as a second randomly selected sample, and in neither of these is the sample statistic likely to be exactly equal to the population parameter it is estimating and about which we are testing a hypothesis.

Because we take a sample, there is a chance of uncertainty involved and, therefore, we should make the conclusions about the population with a probability. This gives an indication of the chance of getting the observed results if the hypothesis is true.

The null hypothesis, *H*0

There are two statistical hypotheses: the null hypothesis (*H*0) and the alternative hypothesis (*H*1). The null hypothesis is usually an assumption of unchanged state. For example, the *H*0 states that there is no difference between some characteristics, for example means or variances of two populations.

The alternative hypothesis, *H*1, describes a changed state or existence of a difference

The research hypothesis can be postulated as two possibilities: there is a **difference or there is no difference. Usually the statistical alternative hypothesis *H*1 is identical to the research hypothesis, thus the null hypothesis is opposite to what a researcher expects.**

It is generally easier to prove a hypothesis is false than that is true, thus a researcher usually attempts to reject *H*0.

A statistical test based on a sample lead to one of two conclusions

1) a decision to reject H0 (because it is found to be false), or
2) a decision not to reject H0, because there is insufficient proof for rejection. The null and alternative hypothesis, H0 and H1, always exclude each other.

Should the null hypothesis be untrue, an alternative hypothesis holds

Usually, the alternative hypothesis states that a difference exists between the parameter values but the direction of that difference is not known. It leads to a two-sided or **a two tailed test**.

Very occasionally, however, we have sound prior knowledge that any difference between the treatments, if it exists, can be in one direction only. This must not be based on hopes or expectations about a novel treatment, but on an absolute certainty that the difference can only be in that direction if the difference is not zero. This gives rise to a **one-sided** or a **one tailed test** in which the direction of the difference is specified in the alternative hypothesis.

The test statistic and the *P*-value

From the data we calculate the value of a **test statistic** (an algebraic expression particular to the hypothesis we are testing), usually using a computer or, occasionally, by hand.

Attached to each value of the test statistic is a **probability**, called a ***P*-value**. It describes the chance of getting the observed effect (or one more extreme) *if the null hypothesis is true*.

The 'if the null hypothesis is true' is crucial to the correct interpretation of the *P*-value; a common mistake is to omit this phrase, leading to the *erroneous* belief that the *P*-value represents the probability of the sample data arising by chance.

Making a decision using the *P*-value

a **decision** whether or not to reject the null hypothesis.

- If, however, the observed results *are* consistent with what we would expect if the null hypothesis were true, we **do not reject** the null hypothesis. We say that the result of the test is **non-significant**.
- If the observed results are *not* consistent with what we would expect *if the null hypothesis were not true*, we conclude that we have enough evidence to **reject** the null hypothesis. We say that the result of the test is **statistically significant**.

Significance level of the test

The level of significance is denoted by the Greek symbol α (alpha). Therefore, the level of significance is defined as follows:

Significance Level = p (type I error) = α

We have to decide, *before we collect the data*, what constitutes a large or small *P*-value, i.e. we have to choose a cut-off value, termed the **significance level** of the test. The choice of significance level is dependent on the nature of the data and the circumstances of the investigation.

It makes sense to choose a very low value for the significance level, say 0.01, if we want to error on the side of caution in rejecting *H*0. This means that if, from our sample, we obtain a *P*-value which is less than 0.01, we reject the null hypothesis; we say the result is *significant at the 1% level.*

An arbitrary cut-off point of 0.05 is usually chosen for the significance level such that if $P < 0.05$ then *H*0 is rejected, and if $P \geq 0.05$ **then *H*0 is not rejected (Fisher, 1925)**.

Further distinctions are sometimes made by using asterisks to distinguish between

(*** representing $P < 0.001$) very highly significant,

(**representing $0.001 < P < 0.01$) highly significant,

(* representing $0.01 < P < 0.05$) significant,

and non-significant (NS representing $P \geq 0.05$).

Degrees of freedom of the test statistic

The degrees of freedom of a statistic are the number of independent observations contributing to that statistic, i.e. the number of observations available to evaluate that statistic minus the number of restrictions on those observations.

So, for example, suppose we are estimating a population variance, σ^2, of a variable, x, in a sample of size n by its sample statistic, s^2, given by

$$s^2 = \frac{\sum (x - \bar{x})^2}{(n-1)}$$

We have to estimate the mean in order to evaluate the numerator, and so the degrees of freedom of s^2 are $(n - 1)$.

The easiest way of calculating the degrees of freedom of any statistic is to take them as the difference between the number of observations we have in our sample and the number of parameters we have to estimate in order to evaluate that statistic.

Quoting a confidence interval

The use of a **confidence interval** diminishes, to some extent, the reliance on the P-value and the subsequent significance or non-significance of a test result. We measure the precision of a sample statistic as an estimate of the **population parameter by its standard error**.

Summary of the hypothesis test procedure

We can assimilate this whole hypothesis test procedure by generalizing it to a six-step procedure:

1. Specify the **null hypothesis**, ***H*0**, and the **alternative hypothesis** (by default, we adopt a two-sided test unless a different alternative hypothesis is specified).
2. **Collect** the data and **look** at them, diagrammatically if possible, to investigate their distribution(s), check for outliers and get a feel for what they show. Many tests make distributional assumptions about the data: check the assumptions underlying the test.

3. Select the **appropriate test**, by hand, calculate the appropriate **test statistic** using the sample data.
4. Relate the calculated value of the test statistic to a ***P*-value**.
5. Consider the *P*-value to judge whether the data are inconsistent with the null hypothesis. Then **decide** whether or not to reject the null hypothesis.
6. If appropriate, calculate the **confidence interval** for the effect of interest, phrased in terms of the parameter specification in the null hypothesis.

Type I and Type II errors

Making the wrong decision in a hypothesis test

1. we have rejected the null hypothesis when we should not have rejected it (i.e. when it is true). We are making a **Type I error.**
2. We have not rejected the null hypothesis when we should have rejected it (i.e. when it is false). We are making a **Type II error.**

	Reject H_0	**Don't reject H_0**
H_0 true	Type I error (α)	Correct decision
H_0 false	Correct decision	Type II error (β)

The **probability of making a Type I error** is the probability of incorrectly rejecting the null hypothesis, i.e. it is the *P*-value obtained from the test. The null hypothesis will be rejected if this probability is less than the significance level, often **denoted by α (alpha)** and commonly taken as 0.05. Thus the **significance level** is the maximum chance of making a Type I error.

If the *P*-value is equal to or greater than α, then we do not reject the null hypothesis and we are not making a Type I error. Therefore, by choosing the significance level of the test to be α at the design stage of the study, we are limiting the probability of a Type I error to be less than α.

The **probability of making a Type II error** is usually designated by β (beta). It is the probability of not rejecting the null hypothesis when the null hypothesis is false. β is affected by a number of factors, one of which is the sample size; the greater the sample size, the smaller β becomes (keeping the other factors that affect it constant).

In fact, instead of thinking about β, we usually consider its complement, $1 - \beta$ (often multiplied by 100 and expressed as a percentage). This is called the **power** of the test. **It is the probability of rejecting the null hypothesis when the null hypothesis is false**, i.e. it is the chance of detecting a treatment effect of a given size if it exists.

Confidence interval approach to hypothesis testing

We then calculate the appropriate (usually 95%) confidence interval for the difference in the two population means. The 95% confidence interval for the difference in two population means is usually interpreted as the interval within which the true difference in means is contained with 95% certainty. It represents the plausible range of values for the difference in population means. Zero is an implausible value for the difference in means if it lies outside the 95% confidence interval.

So, if the confidence interval does not span zero, we would conclude that it is unlikely (the chance is at most 5%) that the difference in population means could be zero, and we would reject the null hypothesis that the means are equal at the 5% level of significance.

To summarize, there are two approaches to testing a null hypothesis

We can calculate the appropriate test statistic and determine the associated P-value. This represents the probability of obtaining the observed value of the test statistic, or one that is more extreme, if the null hypothesis is true. We reject the null hypothesis if the P value is small, say < 0.05. The whole process is called a hypothesis test or a significance test.

We can calculate the 95% confidence interval for the appropriate parameter expression (e.g. the difference in population means, $\mu^1 - \mu^2$, if we are testing H0: $\mu^1 = \mu^2$). We reject the null hypothesis at the 5% level if the value of the parameter expression under the null hypothesis lies outside the confidence limits.

Collecting our thoughts on confidence intervals

1. It is an **indicator of the precision of the parameter estimate**. A wide confidence interval indicates poor precision.
2. **The result of a hypothesis test that two means are equal might be significant but the confidence interval for the difference in the two means could be very narrow and its limits** only just greater than zero. Here we might have statistical significance but the implications may not be biologically important.
3. It can be used to test a null hypothesis about the parameter of interest. For example, the null hypothesis that there is no difference between two treatment means can be rejected, if the confidence interval for the difference excludes zero.

Calculating the confidence interval for the mean

(a) Where σ is known

Provided we have knowledge of σ, the 95% confidence interval for the mean is $x \pm 1.96 \times$ SEM, i.e.

$$\overline{x} \pm 1.96\frac{\sigma}{\sqrt{n}} = \left(\overline{x} - 1.96\frac{\sigma}{\sqrt{n}}, \overline{x} + 1.96\frac{\sigma}{\sqrt{n}}\right)$$

Here, the upper and lower limits of the confidence interval within the bracket are separated by the comma. For the 95% confidence interval, the multiple is 1.96 (often approximated by 2). The multiple is 2.58 for a 99% confidence interval; note that a multiple of 1.00 only gives approximately a 68% confidence interval.

(b) Where σ is unknown

Usually we do not know the value of the population standard deviation, σ, so we replace it by the sample estimate

$$\overline{x} \pm t_{0.05}\frac{s}{\sqrt{n}} = \left(\overline{x} - t_{0.05}\frac{s}{\sqrt{n}}, \overline{x} + t_{0.05}\frac{s}{\sqrt{n}}\right)$$

Example

We summarized the results of the milk yields of a sample of 256 Holstein–Friesian cows; the sample mean was 9414 kg and the estimated SEM was 147 kg. From Table, $t_{0.05}$ for $df = 255$ is approximately 1.96. The 95% confidence interval for the true mean milk yield is given by

$$\begin{aligned}\overline{x} \pm t_{0.05}\frac{s}{\sqrt{n}} &= (9414 - 1.96 \times 147, 9414 + 1.96 \times 147)\\ &= (9125.9, 9702.1)\text{kg}\end{aligned}$$

Hence, we are 95% certain that the mean milk yield for the population of Holstein–Friesian cows lies between 9126 and 9702 kg.

11

Statistical Tests

1. Hypothesis Test of a Population Mean (Z test)

One use of a hypothesis test is to determine if a sample mean is significantly different from population mean. This example of hypothesis testing will be used to show the general principles of statistical hypotheses. First, a researcher must define null and alternative hypotheses. **To determine if a population mean is different from some value $\mu 0$**, the null and alternative hypotheses are:

H_0: $\mu = \mu 0$

H_1: $\mu \neq \mu 0$

The null hypothesis, $H0$, states that the population mean is equal to $u0$, the alternative hypothesis, $H1$, states that the population mean is different from $\mu 0$.

The next step is to define an estimator of the population mean. This is the sample mean $\overline{y}$. Now, a test statistic with a known theoretical distribution is defined. For large samples the sample mean has a normal distribution, so a standard normal variable is defined:

$$z = \frac{\overline{y} - \mu_o}{\sigma_{\overline{y}}}$$

Where,

$$\sigma_{\overline{y}} = \frac{\sigma}{\sqrt{n}}$$

So, Z statistics has a normal distribution when $\mu = \mu 0$, that is if H_0 is correct.

Z= (Estimator – Parameter)/ standard error of the estimator

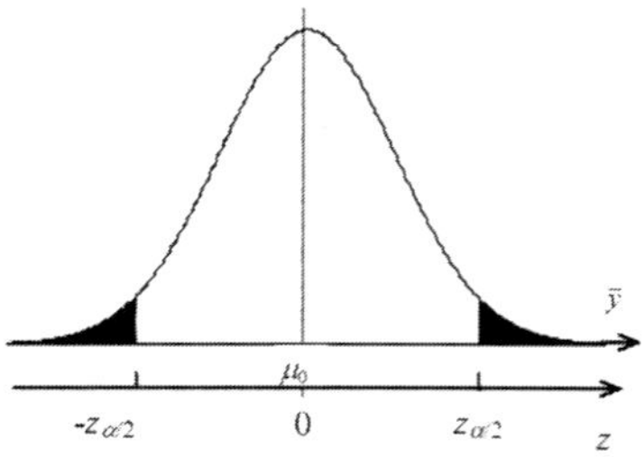

Fig. 1: Distribution of $\bar{y}$, lower scale is the standard scale

$$z = \frac{\bar{y} - \mu_0}{\sigma_{\bar{y}}}$$

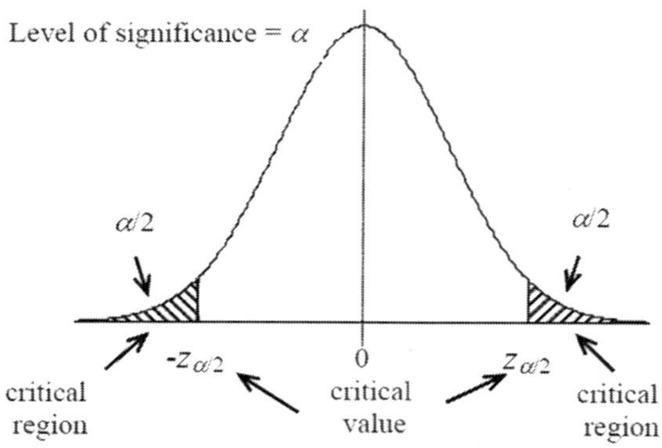

Fig. 2: Illustration of significance level critical value and critical region

Example 1: Given a sample of 50, cows with an arithmetic mean for lactation milk yield of 4000 kg, does this herd belong to a population with a mean μ0 = 3600 kg and a standard deviation σ = 1000 kg?

The hypothetical population mean is μ0 = 3600 and the hypotheses are:

$H_0 : \mu = 3600$

$H_1 : \mu \neq 3600$

Known values are:

$\bar{Y} = 4000$ kg

σ = 1000 kg

n = 50 cows

The calculated value of the standard normal variable is:

$Z=(4000-3600)/(1000/\sqrt{50})=2.828$

A significance level of $\alpha = 0.05$ is chosen. The critical value corresponding to $\alpha = 0.05$ is $z\alpha/2 = 1.96$. The calculated z is 2.828.

The question is if the calculated $z = 2.828$ is extreme enough that the sample does belong to the population with a mean 3600 kg. The calculated $|z| > z\alpha/2$, numerically, $|2.828| > 1.96$, which means that the calculated z is in the critical region for rejection of H0 with $\alpha = 0.05$ level of significance (Figure 3).

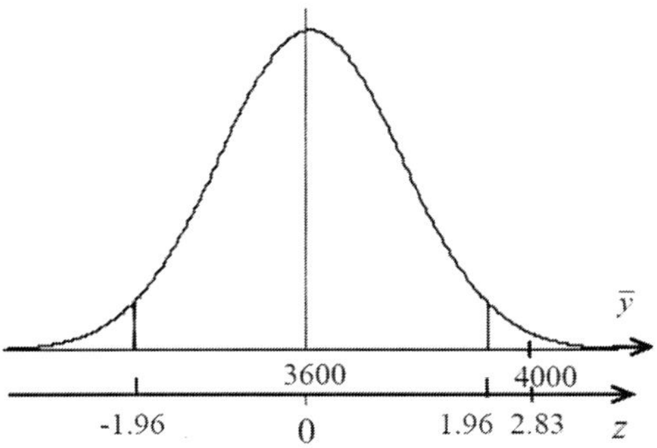

Fig. 3: A distribution of sample belongs to the population with the mean of 3600 kg and standard deviation of 1000. The lower line represents the standard normal scale.

Example 2: In an experiment on two sets of 40 and 50 animals, two diets were compared. the average increase in weights due to two diets A and B was 5 kg and 6 kg, respectively and population's standard deviation for increase in weights is 1.2 kg. Test whether there are significant differences in weight gain due to diets.

Answer

1. $H_0 : \bar{X}_1 = \bar{X}_2$ i.e. Average increase in weights of two sets of animals are equal.

   $H_A : \bar{X}_1 \neq \bar{X}_2$ i.e. Average increase in weights of two sets of animals are not equal.

2. Level of significance is taken either 5% or 1%. We fix the level significance at 5% and critical value of $z = 1.96$ at 5% level.

3. **Assumption:** Sample should be drawn randomly.

   Population should be normally distribution.

   Population standard deviation should be known.

4. **Calculation**

   $n_1 = 40$ $n_2 = 50$ $\overline{Y}_1 = 5$ kg $\overline{\ }$ $= 6$ kg $\sigma = 1.2$

   $$Z = \frac{\overline{Y}_1 - \overline{Y}_2}{\sqrt{\sigma^2}(1/n_1 + 1/n_2)}$$

   z = (5-6)/0.2545 =- 3.92

5. **Conclusion:** Absolute calculated value of Z is more than tabulated value. The negative sign shows that Z lies in left critical region and null hypothesis is rejected. It indicates that alternate hypothesis is accepted i.e. Diet A is inferior than diet B.

Example 3

1. The following data give the mean and standard deviations for the 6 months body weights (kg) of groups of Rathi and Crossbred calves.

	Mean weights	SD	Sample size
Rathi calves	100.8	4.5	200
Crossbred calves	102.5	1.5	300

Test the hypothesis that means weights do not differ between breeds.

Answer

1. $H_0 : \overline{X}_1 = \overline{X}_2$ i.e. means weights do not differ between breeds.

   $H_A : \overline{X}_1 \neq \overline{X}_2$ i.e. means weights differ between breeds.

   Level of significance is taken either 5% or 1%. We fix the level significance at 5% and critical value of z = 1.96 at 5% level.

2. **Assumption:** Sample should be drawn randomly.

   Population should be normally distribution.

   Population standard deviation should be known.

3. **Calculation**

   If the variances are unknown, they can be estimated from the samples and the standard error is:

   $$S_{(\bar{y}_1 - \bar{y}_2)} = \sqrt{\frac{S_1^2}{n_1} + \frac{S_2^2}{n_2}}$$

So S^2_1 and S^2_2 are the estimate of variance of the two samples. Then the Z statistics is:

$$z \approx \frac{\overline{y}_1 - \overline{y}_2}{S_{(\overline{y}_1 - \overline{y}_2)}}$$

Y1-Y2=100.8-102.5= -1.7

SE= SQRT OF 4.5*4.5/200 + 1.5*1.5/300

= SQRT OF 0.10875

= 0.329

Z = -1.7/0.329 = -5.16

5. **Conclusion:** Calculated value is higher than table value so Ho is rejected, means test is significant and H1 is accepted. Negative z value indicates Y2 means crossbred calves are better body wight than Rathi calves.

Student's t-distribution

In 1908 William Sealy Gosset, an Englishman publishing under the pseudonym student, developed the t-test and t distribution. Gosset worked at the Guinness brewery in Dublin and found that existing statistical techniques using large samples were not useful for the small sample sizes that he encountered in his work.

2. Hypothesis Test of a Population Mean for a Small Sample

The student's t-distribution is used for testing hypotheses about the mean for a small sample (say n < 30) drawn from a normal population. The test statistic is a t random variable:

$$t = \frac{\overline{y} - \mu_0}{s / \sqrt{n}}$$

The approach to reaching a conclusion is similar to that for a large sample. The calculated value of the t statistic is tested to determine if it is more extreme than the critical value tα or tα/2 with α level of significance.

For a two-sided test the null hypothesis H0: μ = μ0 is rejected if |t| > tα/2, where tα/2 is a critical value such that P(t > tα/2) = α/2. For a one-sided test the null hypothesis is rejected if t > tα or H0: μ ≥ μ0 is rejected if t < –tα, depending on whether it is a right or left. Critical Values of the Student *t*-distribution in Appendix. The shape of the distribution and the value of the critical point depend on degrees of freedom. The degrees of freedom are (*n* – 1), where *n* is the number of observations in a sample.

Example 1: Observations of sample are 5, 3, 1, 4, 2. Test with 5% significant level that the population mean is 5.

Solution

1. H_0 : Sample belong to hypothetical population with mean 5.

   H_A : Sample does not belong to hypothetical population.

2. Level of significance 5% Critical value of 't' at 4 degrees of freedom = 2.77.

3. **Assumption**: Sample should be drawn randomly.

   Population should be normally distributed.

4. **Calculation:**

   $\overline{Y} = \Sigma Y/N = 15\sqrt{5} = 3$

   $\Sigma Y^2 = 55, \Sigma Y = 15, n = 5$

   $$S^2 = \frac{55-(15)^2}{5-1} = 2.5$$

   $SE = \sqrt{2.5/5} = 0.7071$

   $$t = \frac{Y\text{-}\mu}{SE\ of\ \overline{Y}} = \frac{3-5}{0.7071} = -2.83$$

Conclusion

Absolute calculated value of 't' is more than 2.77 and lie in critical region. Therefore, H_0 is rejected. Negative sign shown that sample belong to a population, which has mean less than 5.

Example 2: The data are lactation milk yields of 10 cows. Is the arithmetic mean of the sample, 3800 kg, significantly different from 4000 kg? The sample standard deviation is 500 kg.

The hypothetical mean is $\mu 0 = 4000$ kg and the hypotheses are as follows:

$H0$: $\mu = 4000$ kg

$H1$: $\mu \neq 4000$ kg

sided test. Critical values can be found in the table

The sample mean is $\overline{Y} = 3800$ kg.

The sample standard deviation is s = 500 kg.

The standard error is:

$$s/\sqrt{n} = 500/\sqrt{10} \quad t = \frac{\bar{y} - \mu_0}{s/\sqrt{n}} = \frac{3800 - 4000}{500/\sqrt{10}} = -1.26$$

For α = 0.05 and degrees of freedom (n – 1) = 9, the critical value is –tα/2 = –2.262. Since the calculated t = –1.26 is not more extreme than the critical value –tα/2 = –2.262, H0 is not rejected with an α = 0.05 level of significance. The sample mean is not significantly different from 4000 kg.

3. Hypothesis Test of the Difference between Two Population Means

Assume that samples are drawn from two populations with means $\mu 1$ and $\mu 2$. The samples can be used to test if the two means are different. The z or t statistic will be used depending on the sample size. The form of the test also depends on whether the two samples are dependent or independent of each other, and whether the variances are equal or unequal.

Further, the hypotheses can be stated as one or two-sided.

The hypotheses for the two sided test are:

$H0: \mu 1 - \mu 2 = 0$

$H1: \mu 1 - \mu 2 \neq 0$

The null hypothesis $H0$ states that the population means are equal, and the alternative

hypothesis $H1$ states that they are different.

$$z = \frac{(\bar{y}_1 - \bar{y}_2) - 0}{\sigma_{(\bar{y}_1 - \bar{y}_2)}}$$

where

$$\sigma_{(\bar{y}_1 - \bar{y}_2)} = \sqrt{\frac{\sigma_1^2}{n_1} + \frac{\sigma_2^2}{n_2}}$$

is the standard error of difference between the two means and σ^2_1 and σ^2_2 are the variances of the two populations.

If the variances are unknown, they can be estimated from the samples and the standard error is:

$$S_{(\bar{y}_1 - \bar{y}_2)} = \sqrt{\frac{S_1^2}{n_1} + \frac{S_2^2}{n_2}}$$

where S^2_1 and S^2_2 are estimates of the variances of the sample.

Then the z statistics is,

$$z \approx \frac{\bar{y}_1 - \bar{y}_2}{S(\bar{y}_1 - \bar{y}_2)}$$

For a two-sided test H_0 is rejected if the calculated value $|z| > z\alpha/2$, where $z\alpha/2$ is the critical value for the significance level α.

In order to reject H_0, the calculated value of the test statistic z must be more extreme than the critical value $z\alpha/2$.

Example: Two group of 40 cows were fed two different rations (A and B) to determine which of those two rations will yield more milk in lactation. At the end of the experiment the following sample means and variances (in thousand kg) were calculated:

	Ration A	Ration B
Mean ($\bar{Y}$)	5.20	6.50
Variance (s^2)	0.25	0.36
Size (n)	40	40

The hypothesis for a two-sided test are:

H0: μ1= μ2

H1: μ1≠ μ2

The standard error of difference is:

$$S_{(\bar{y}_1 - \bar{y}_2)} = \sqrt{\frac{S_1^2}{n_1} + \frac{S_2^2}{n_2}} = \sqrt{\frac{0.25}{40} + \frac{0.36}{40}} = 0.123$$

The calculated value of Z statistics is:

$$z \approx \frac{\bar{y}1 - \bar{y}2}{S(\bar{y}1 - \bar{y}2)} = \frac{5.20 - 6.50}{0.123} = -10.569$$

Since the calculated value z= -10.569 is more extreme than the critical value -1.96, the null hypothesis is rejected with α = 0.05 level of significance, suggesting that feeding cows ration B will results in greater milk yields than ration A.

4. Small Samples and Equal Variances

For comparisons involving small samples a t statistic is used. The definition of the t statistic depends on whether variances are equal or unequal. The test statistic for small samples with equal variance is:

$$t=\frac{(\bar{y}_1-\bar{y}_2)}{\sqrt{s_p^2\left(\frac{1}{n_1}+\frac{1}{n_2}\right)}}$$

where y_1 and y_2 are the sample means, n_1 and n_2 are sample sizes, and s^2p is the pooled variance:

$$S_p^2=\frac{(n_1-1)S_1^2+(n_2-1)S_2^2}{n_1+n_2-2}$$

Here $i = 1$ to $n1$, $j = 1$ to $n2$. Since the variances are assumed to be equal, the estimate of the pooled variance $s^2\ p$ is calculated from the observations from both samples.

When the number of observations is equal in two samples, that is $n1 = n2 = n$, the expression for the t statistic simplifies to:

$$t=\frac{(\bar{y}_1-\bar{y}_2)}{\sqrt{\frac{S_1^2+S_2^2}{n}}}$$

Example: The litter size at birth in two breed of rabbit is recorded as follows:

Breed A:	4	5	4	6	3
Breed B:	6	7	6	8	9

Test with 5% significant level that the two samples belongs to the same population.

Solution

1. $H_0 : \bar{X} = \bar{Y}$

   $H_A : \bar{X} \neq \bar{Y}$
2. Assumption: Sample should be drawn randomly.

   Population should be normally distributed.
3. Level of significance: The tabulated value of t at 5% of significance and 8 degrees of freedom is 2.3.
4. Calculation:

   Mean for breed A: 22/5 = 4.4 for breed B = 36 / 5 = 7.2

   $\sum A = 22$ $\sum A^2 = 102$ $\sum B = 36$ $\sum B^2 = 266$

   $S_A{}^2 = 1.3$ $S_B{}^2 = 1.7$ Pooled variance = 1.5

SE = 0.547

t = (7.2- 4.4) / 0.547 = 2.8/0.547 = 5.118

5. **Conclusion:** Table value of 't' at 8 df at 5% level of significance is 2.3 that is lower than calculated value. Hence, difference is significant and null hypothesis is rejected and two samples are not same. They belong to different population.

5. Small Sample and Unequal Variance

A statistics for testing the difference between two population means with unequal variances is also a t statistics:

$$t = \frac{(\bar{y}_1 - \bar{y}_2) - 0}{\sqrt{\left(\frac{S_1^2}{n_1} + \frac{S_2^2}{n_2}\right)}}$$

For unequal variances degrees of freedom, denoted v, are no longer equal to (n1 + n2-2) but are:

$$v = \frac{(S_1^2 / n_1 + S_2^2 / n_2)^2}{\frac{(S_1^2 / n_1)^2}{n_1 - 1} + \frac{(S_2^2 / n_2)^2}{n_2 - 1}}$$

6. Dependant sample (Paired t test)

Under some circumstances two samples are not independent of each other. A typical example is taking measurements on the same animal before and after applying a treatment. The effect of the treatment can be thought of as the average difference between the two measurements. The value of the second measurement is related to or depends on the value of the first measurement. In that case the difference between measurements before and after the treatment for each animal is calculated and the mean of those differences is tested to determine if it is significantly different from zero. Let *di* denote the difference for an animal *i*. The test statistic for dependent samples is:

$$t = \frac{\bar{d} - 0}{s_d / \sqrt{n}}$$

where $\bar{d}$ and s_d are the mean and standard deviation of the differences, and n is the number of animals.

The testing procedure and definition of critical values is as before, except that degrees of freedom are $(n - 1)$. For this test to be valid the distribution of observations must be approximately normal.

Example 1: Forced Expiratory Volume (FEV) on 5 asthmatics patients before and after the use of a bronchodilator (litres/second) is given below.:

Before (X)	1.5	1.7	2.1	1.6	2.4
After (Y)	1.7	1.9	2.2	1.9	2.4

Test whether there is any change in the mean FEV value after the used of a bronchodilator

Solution

1. **Hypothesis:**

   $H_0 : \bar{X} = \bar{Y}$

   $H_A : \bar{X} \neq \bar{Y}$

2. **Assumption:** a. Sample should be drawn randomly.

   b. Population should be normally distributed.

3. **Level of significance:** The tabulated value of t at 5% of significance and 4 degrees of freedom is 2.77.

4. **Calculation**

	$(X_1\text{-}Y_1)$	$(X_2\text{-}Y_2)$	$(X_3\text{-}Y_3)$	$(X_4\text{-}Y_4)$	$(X_5\text{-}Y_5)$	**Total**
D	-0.2	-0.2	-0.1	-0.3	0	-0.8
D^2	0.04	0.04	0.01	0.09	0	0.18

$\bar{D}$ = - 0.8/5 = - 0.16

Variance = $(\sum D^2 - (\sum D)^2 / n) / (n - 1)$

= (0.18 – 0.128) / 4

= 0.013

SE = $\sqrt{0.013/5}$

= 0.051

t = - 0.16/ 0.051 = - 3.14

5. Conclusion

The absolute value of calculated 't' – test is more than table value indicating significant difference between forced expiratory volume before and after the use of bronchodilator. Negative sing indicate that FEV is increased significantly more after the use of bronchodilator than before the use of bronchodilator.

Example 2: To study the effectiveness of a new drug in reducing cholesterol level, serum cholesterol determinations in (mg/100 ml) of six patients before and after the administration of drug were found as follows:

Before: 212 250 235 200 215 220

After: 215 245 220 225 210 215

Do the data provide sufficient evidence to conclude that the drug is effective in reducing serum cholesterol level?

Solution:

1. **Hypothesis:**

   $H_0 : \overline{X} = \overline{Y}$

   $H_A : \overline{X} \neq \overline{}$

2. **Assumption:** a. Sample should be drawn randomly.

   b. Population should be normally distributed.

3. **Level of significance:** The tabulated value of t at 5% of significance and 5 degrees of freedom is 2.015.

4. **Calculation**

Sl. No.	1	2	3	4	5	6			
X	212	250	235	200	215	220			
Y	215	245	220	225	210	215	SUM	AVG	VARIANCE
D=X-Y	-3	5	15	-25	5	5	2	0.33	186.6667
D^2	9	25	225	625	25	25	934		
VARIANCE		186.668							
	SD	13.7							
	SE	5.6							
	t	0.0589							

5. **Conclusion:** t table value of t is 2.015 at 5 df, As calculated value of t is less than table vale, H_0 is accepted, test is non-significant. So new drug has no effect on reducing cholesterol level.

Example 3: The effect of a treatment is tested on milk production of dairy cows. The cows were in the same parity and stage of lactation. The milk yields were measured before and after administration of the treatment:

Measurement	Cow1	Cow2	Cow3	Cow4	Cow5	Cow6	Cow7	Cow8	Cow9
1	27	45	38	20	22	50	40	33	18
2	31	54	43	28	21	49	41	34	20
Difference(d)	4	9	5	8	-1	-1	1	1	2

$n = 9$

$$\bar{d} = \frac{\sum_i d_i}{n} = \frac{4 + 9 + ... + 2}{9} = 3.11$$

$$s_d = \sqrt{\frac{\sum_i (y_i - \mu)^2}{n-1}} = \sqrt{\frac{\sum_i y_i^2 - \frac{\left(\sum_i y_i\right)^2}{n}}{n-1}} = 3.655$$

$$t = \frac{\bar{d} - 0}{s_d / \sqrt{n}} = \frac{3.11 - 0}{3.655 / \sqrt{9}} = 2.553$$

The critical value for $(n - 1) = 8$ degrees of freedom is $t0.05 = 2.306$. Since the calculated value $t = 2.553$ is more extreme than the critical value 2.306, the null hypothesis is rejected with $\alpha = 0.05$ level of significance. The treatment thus influences milk yield.

7. Hypothesis Test of a Sample proportion with Population Proportion

Recall that a proportion is the probability of a successful trial in a binomial experiment. For a sample of size n and a number of successes y, the proportion is equal to: p=y/n

The two sided test for a large sample has the following hypotheses:

$H_0: p = p_0$

$H_1: p \neq p_0$

A z random variable is used as a test statistic:

$$z = \frac{\hat{p} - p_0}{\sqrt{p_0 q_0 / n}}$$

Where,

$\hat{q} = 1 - \hat{p}$

8. Hypothesis Test of the Difference between Proportions from Two Populations

Let y_1 and y_2 be the number of successes in two binomial experiments with sample sizes n_1 and n_2, respectively. For the estimation of $p_1 - p_2$, where p_1 and p_2 are the proportions of successes in two populations, proportions ^p_1 and ^p_2 from two samples can be used:

$$\hat{p}_1 = \frac{y_1}{n_1} \text{ and } \hat{p}_2 = \frac{y_2}{n_2}$$

An estimator of the difference between proportions is

$$\hat{p}_1 - \hat{p}_2$$

The estimator has variance:

$$\frac{p_1 q_1}{n_1} + \frac{p_2 q_2}{n_2}$$

Where q_1=(1-p_1) and q_2= (1-p_2). The hypothesis for a two-sided test are:

H_0: p_0=p_1

H_1: $p_0 \neq p_1$

Where $S_{\hat{p}_1 - \hat{p}_2}$ is the standard error of the estimated difference between proportion $(\hat{p}_1 - \hat{p}_2)$. Since the null hypothesis is that the proportions are equal then:

$$S_{\hat{p}_1 - \hat{p}_2} = \sqrt{\frac{\hat{p}\hat{q}}{n_1} + \frac{\hat{p}\hat{q}}{n_2}}$$

Or,

$$S_{\hat{p}_1 - \hat{p}_2} = \sqrt{\hat{p}\hat{q}\left(\frac{1}{n_1} + \frac{1}{n_2}\right)}$$

Where,

$$\hat{q} = 1 - \hat{p}$$

The proportion $\hat{p}$ is an estimator of the total proportion based on both samples:

$$\hat{p} = \frac{y_1 + y_2}{n_1 + n_2}$$

From the given sample proportions, the estimate of the total proportion can be calculated:

$$\hat{p} = \frac{\hat{p}_1 n_1 + \hat{p}_2 n_2}{n_1 + n_2}$$

From this,

$$z = \frac{(\hat{p}_1 - \hat{p}_2) - 0}{\sqrt{\hat{p}\hat{q}\left(\frac{1}{n_1} + \frac{1}{n_2}\right)}}$$

9. Chi-square Test of goodness of fit of the Difference between Observed and Expected Frequencies

Test was developed by Karl Pearson in 1900.

Assume for some categorical characteristic the number of individuals in each of k categories has been counted. A common problem is to determine if the numbers in the categories are significantly different from hypothetical numbers defined by the theoretical proportions in populations:

H_0: $p1 = p1,0, p2 = p2,0, ..., pk = pk,0$

(that is $H0$: $pi = pi,0$ for each i)

H_1: $pi \neq pi,0$ for at least one i

Where

$$p_i = \frac{y_i}{n}$$

is the proportion in any category i, and $pi,0$ is the expected proportion, n is the total number of observations, $n = \Sigma i\ yi$, $i = 1,..., k$.

A test statistic

$$\chi^2 = \sum_i \frac{[y_i - E(y_i)]^2}{E(y_i)}$$

or,

$$Test = \sum \frac{(O - E)^2}{E}$$

Where,

O = Observed value

E= Expected value

has a *chi*-square distribution with $(k - 1)$ degrees of freedom. Here, k is the number of categories, and $E(yi) = n\,pi,0$ is the expected number of observations in category i.

The null hypothesis, H_0, is rejected if the calculated $\chi 2 > \chi 2\alpha$, where $\chi 2\alpha$ is a critical value for the significance level α, that is a value of $\chi 2$ such that $P(\chi 2 > \chi 2\alpha) = \alpha$.

This holds when the samples are large enough, usually defined as when the expected number of observations in each category is greater than five.

10. Comparing two proportions in a 2 × 2 table using the Chi-squared test

Thus, we can compare the two proportions by investigating the association between the two factors that define the contingency table.

Our null hypothesis is that there is no association between the two factors (outcome, group), equivalent to the null hypothesis that the two population proportions are equal.

Observed frequency

Outcome	Group 1	Group 2	Row marginal total
Success	a	b	a+b
Failure	c	d	c+d
Column marginal total	n1=a+c	n2=b+d	n=a+b+c+d
Observed proportion of success	p1=a/(a+c)	P2=b/(b+d)	P=(a+b)/(a+b+c+d)

Expected frequency:

Outcome	Group 1	Group 2
Success	(a+c)(a+b)/n	(b+d)(a+b)/n
Failure	(a+c)(c+d)/n	(b+d)(c+d)/n

$$Test = \frac{n(|ad - bc| - 0.5n)^2}{(a+b)(c+d)(a+c)(b+d)}$$

The 0.5 in the numerator is known as Yates's correction, a continuity correction included to remove bias, when expected frequency less than 5.

Obtain a P-value, usually from the computer output, but you can derive it by referring the calculated value of the test statistic to the table of the Chi-squared distribution. The test statistic has 1 degree of freedom, equivalent to the number of rows minus 1 (i.e. 2 – 1 = 1) times the number of columns minus 1 (i.e. 2 – 1 = 1).

Problem 1:

A dice is thrown 120 times and frequency of coming different faces is obtained as follows:

Face No:	1	2	3	4	5	6
Frequency:	17	22	25	15	18	23

Solution

1. **Hypothesis:**

   H_0 : Observed frequency = Expected frequency

   H_1 : Observed frequency ≠ Expected frequency

2. **Assumption:**

   Sample should me drwan randomly.

   Population should be randomly distributed.

3. Chi square table value at 5 % level of significance, 5 degree of freedom is 11.07

4. **Calculation:**

Face No	1	2	3	4	5	6	
Observed Frequency (O)	17	22	25	15	18	23	
Expected Frequency (E)	20	20	20	20	20	20	
(O-E)	-3	2	5	-5	-2	3	
(O-E)	9	4	25	25	4	9	Sum
$(O-E)^2$ /E	0.45	0.2	1.25	1.25	0.2	0.45	3.8
Chi Square Calculate Value	3.8		Table value 11.07				

5. **Conclusion:** Chi square calculated value is 3.8, less than table value at 5 % level of significance, 5 degree of freedom 11.070 , so H0 is accepted, test is non significant.

Problem 2

The following table shows the results of inoculation against anthrax in goats.

	Survived	Died
Inoculated	160	60
Not inoculated	80	60

Is vaccine effective against anthrax?

1. **Hypothesis:**

   H_0 : Observed frequency = Expected frequency

   H_1 : Observed frequency ≠ Expected frequency

2. **Assumption:**

   Sample should me drwan randomly.

   Population should be randomly distributed.

3. Chi square table value at 5 % level of significance, 1 degree of freedom is 3.841 and 6.635 at 1 % level.

4. **Calculation**

	Survived	Died	
Inoculated	160 (a)	60(b)	R1=a+b=220
Not Inoculatd	80(c)	60(d)	R2=c+d=140
	C1=a+c=240	C2=b+d=120	
N=240+120=360			
Chi square value= [360x(160x60-80x60}]²/ (240x120x220x140)=2880/308=9.35			
Chi square table value= 3.84 at 0.05% level			
Chi square table value= 6.635 at 0.01% level			

5. Conclusion

Chi square calculated value 9.35 is higher than table value at both 0.05% and 0.01% level, 3.84 and 6.635, So test is highly significant, Ho rejected, H1 accepted. Vaccine is effective against Anthrax vaccine.

Summary

What is Type I error and Type II error?

Answer:

a) **Type I error = rejection of H_0, when H_0 is acually true**

b) **Type II error = failure to reject H_0, when H_0 is actually false**

Decision of a statistical test	H_0 correct No true differnce	H_0 not correct/false a true differnce exist
H0 not rejected/accepted	**Correct acceptance** **P=1- α**	**Type II error** **P=β**
H0 rejected	**Type I error** **P= α**	**Correct rejection** **P= 1- β**

The following have influence on making a correct conclusion:

1. Sample size
2. Level of significance α
3. Effect size (desired difference considering variablity)
4. Power of test (1- β)

11. Coefficient of Correlation and Tests of Hypotheses

To test the significance of a correlation in a sample, the null and alternative hypotheses

about the parameter ρ are:

$H_0 : \rho = 0$

$H_1 : \rho \neq 0$

The null hypothesis states that the coefficient of correlation in the population is not different from zero, that is, there is no linear association between variables in the population. The alternative hypothesis states that the correlation in the population differs from zero. In hypothesis testing, a t-distribution can be used, because it can be shown that the t statistic:

$$t = \frac{r}{s_r}$$

has a t-distribution with $(n - 2)$ degrees of freedom assuming the following:

1) variables x and y have a joint bivariate normal distribution
2) the hypothesis $H_0: \rho = 0$ is true.

$$s_r = \sqrt{\frac{1-r^2}{n-2}}$$

is the standard error of the coefficient of correlation. Further

$$t = \frac{r-0}{\sqrt{\frac{1-r^2}{n-2}}}$$

Or simplified,

$$t = \frac{r\sqrt{n-2}}{\sqrt{1-r^2}}$$

12. Student *t* test in Testing Hypotheses about the Parameters (Regression coefficients)

If changing variable x effects change in variable y, then the regression line has a slope, that is, the parameter $\beta 1$ is different from zero. To test if there is a regression in a population the following hypotheses about the parameter $\beta 1$ are stated:

$H_0 : \beta 1 = 0$

$H_1 : \beta 1 \neq 0$

The null hypothesis $H0$ states that slope of the regression line is not different from zero and that there is no linear association between the variables. The alternative hypothesis $H1$ states that the regression line is not horizontal and there is linear association between the variables. Assuming that the dependent variable y is normally distributed, the hypotheses about the parameter $\beta 1$ can be tested using a t distribution. It can be proved that the test statistic:

$$t = \frac{b_1 - 0}{\sqrt{S^2 / SS_{xx}}}$$

has a t distribution with $(n - 2)$ degrees of freedom under $H0$. Note the form of the t statistic:

t =Estimator- Paramater (under H0)/ Standard error of estimator

The null hypothesis $H0$ is rejected if the computed value from a sample $|t|$ is "large". For a level of significance α, $H0$ is rejected if $|t| \geq t\alpha/2,(n\text{-}2)$, where $t\alpha/2,(n\text{-}2)$ is a critical value.

13. F test for Partitioning Total Variability and Tests of Hypotheses (Regression coefficients):

As for the simple regression model, the total sum of squares can be partitioned into

regression and residual sum of squares:

$SS_{TOT} = SS_{REG} + SS_{RES}$

The regression sum of squares is:

$= (\hat{y} - \overline{y})'(\hat{y} - \overline{y}) = \sum_i (\hat{y}_i - \overline{y})^2$

The residual sum of squares is:

$SS_{RES} = (y - \hat{y})'(y - \hat{y}) = \sum_i (y_i - \hat{y}_i)^2$

The total sum of squares is:

$= (y - \overline{y})'(y - \overline{y}) = \sum_i (y_i - y)^2$

or shortly, using the computed vector **b**:

$SS_{REG} = b'X'y - n\overline{y}^2$

$SS_{RES} = y'y - b'X'y$

$SS_{TOT} = y'y - n\overline{y}^2$

The null and alternative hypotheses are:

H_0: $\beta 1 = \beta 2 = 0$

H_1: at least one $\beta i \neq 0$, $i = 1$ and 2

If $H0$ holds then the statistic:

$F = MS_{REG} / MS_{RES}$

has an F distribution with $(p - 1)$ and $(n - p)$ degrees of freedom. For the α level of significance H_0 is rejected if the calculated F is greater than the critical value $F\alpha, p\text{-}1, n\text{-}p$.

The *ANOVA* table is:

Source	SS	df	MS=SS/df	F
Regression	SS_{REG}	p - 1	$MS_{REG} = SS_{REG}/(p-1)$	$F = MS_{REG}/MS_{RES}$
Residual	SS_{RES}	n - p	$MS_{RES} = SS_{RES}/(n-p)$	
Total	SS_{TOT}	n - 1		

The coefficient of multiple determination is:

$$R^2 = \frac{SS_{REG}}{SS_{TOT}} = 1 - \frac{SS_{RES}}{SS_{TOT}}$$

The value of R^2 ranges from 0 to 1.

$0 \leq R^2 \leq 1$

12

F Test

The F-test for the equality of two variances

Rationale

The two-sample t-test and the analysis of variance make the assumption of *homoscedasticity*, i.e. of equal variances in groups of data. The F-test, often called the variance ratio test, may be used to investigate the homoscedasticity of two data sets.

We have to make a decision whether or not the population variances are likely to be different. This means that we need a cut-off for the variance ratio; if the variance ratio exceeds this cut-off value, we will conclude that the variances are unequal.

We determine this cut-off value formally, under the *null hypothesis* that the two population variances are equal, by referring the ratio to the table of the F-distribution. The *degrees of freedom* are $(n_1 - 1)$ in the numerator (the larger variance) and $(n_2 - 1)$ in the denominator (the smaller variance), where n_1 and n_2 represent the two sample sizes.

For the required significance level in a two-tailed test, therefore, we must halve this tail area. So, for a two-tailed test at the 5% level of significance, we have to relate the test statistic to $P = 0.025$. For convenience, we give the upper percentage points corresponding to $P = 0.025$ and $P = 0.005$ (relating to two-tailed P-values of 0.05 and 0.01, respectively) in separate tables.

Assumptions

This hypothesis test is dependent on the assumptions that the independent samples are selected from Normally distributed populations and are representative of these populations (ideally having been chosen by random selection). It is particularly sensitive to departures from normality.

Approach

1. Specify the null hypothesis that the two population variances, σ_1^2 and σ_2^2, are equal, and specify the alternative hypothesis, usually that the variances are not equal.
2. Collect and examine the data by constructing a dot plot, a histogram or a box-and-whisker plot. Check the Normality assumption.
3. Select the appropriate test on the computer, or calculate by hand the test statistic. This is the $F= s_1^2 / s_2^2$, ratio of the larger (s_1^2) to the smaller (s_2^2) estimated population variances derived from the samples. It follows the F-distribution and is given by Test with $n_1 - 1$, $n_2 - 1$ degrees of freedom.
4. Determine the ***P*-value**. You may find this in the computer output or, alternatively, turn to the *F*-table. You will see that F-Table in Appendix, have both columns and rows that correspond to the degrees of freedom for the numerator and the denominator, respectively, of the *F*-ratio.
5. Use the *P*-value to judge whether the data are inconsistent with the null hypothesis. Then **decide** whether or not to reject the null hypothesis. Commonly, we reject the null hypothesis if $P < 0.05$.

Analysis of variance (ANOVA) for the equality of means

Rationale

a) The analysis of variance (ANOVA) is an expression used to describe a set of techniques that compare the means of two, three or more groups by investigating relevant variances.

b) The analysis is based on the variance ratio test, i.e. an F-test, which compares two variances by examining their ratio and relating it to the F-distribution'

c) The principle underlying the analysis of variance is that the total variability in a data set is partitioned into its component parts.

d) Each component represents a different source of variation.

e) We may regard the one-way ANOVA (the simplest form of ANOVA) as an extension of the two-sample t-test when we compare the means of more than two groups. They give identical P-values when there are two groups.

f) The one-way repeated measures ANOVA may be regarded as an extension of the paired t-test, for example when the data are longitudinal and are measured on each individual at successive time points, or when the within-subject/ animal comparison is between three or more treatments.

g) The two-way ANOVA examines the effect of two factors on a response variable, when each of these factors possesses two or more levels.

h) We create a two-way table in which the rows and the columns represent the levels of each of the two factors, every cell in the table represents a unique combination of particular levels of the two factors.

Providing there is no replication in any cell of the table, the design is often called a **randomized block**. If there is replication in the cells, then it is possible to study the **interaction** between the factors.

One-way analysis of variance Assumptions

a) The assumptions of the one-way ANOVA are that the samples representing the levels are independent and the observations in each sample come from a normally distributed population with variance σ^2.

b) This implies that the group variances are the same. Approximate Normality may be established by drawing a histogram; moderate departures from normality have little effect on the result.

Approach

1. Specify the **null hypothesis** that the population group means do not differ, i.e. the groups represent a single population. Generally, the **alternative hypothesis** is that at least one of the group means is dissimilar.
2. **Collect** the data and **display** them in exactly the same way as for the **two-sample *t*-test** except that here are more than two groups. Check the assumptions of normality and homogeneity of variance.
3. Use the P-value to judge whether the data are inconsistent with the null hypothesis. Then decide whether or not to reject the null hypothesis; usually, but not necessarily, we reject H0 if $P < 0.05$. If we reject the null hypothesis, we may need to establish which group means differ.
4. Derive the **confidence intervals for differences between group means**, in essentially the same way as for the two-sample t-test. However, the combined estimate of variance used in the calculation of the confidence interval is the within-group (residual) mean square in the ANOVA table.

The best way of establishing whether the estimate is good is to calculate what is called the **confidence interval for the mean**. This interval, defined by its upper and lower limits (the **confidence limits**), is generally interpreted as the range of values within which we expect the true population mean to lie with a certain probability.

If F is significant or highly significant, critical difference between treatment means is to be worked out

Critical difference between any two-treatment means is defined as the **least significant difference** between any two treatment means, to be exceeded by the difference between two treatment means to declare them as significantly different.

Critical difference between any two treatment means at 5% level = Standard error of the difference between the treatment means x table value of ‘t’ for error d.f. at 5% level.

Critical difference between any two treatment means at 1% level = Standard error of the difference between the treatment means x table value of ‘t’ for error d.f. at 1% level.

Critical difference between Tr1 and Tr2 at 5% (1%)

$$\sqrt{EMS\left(\frac{1}{n_1}+\frac{1}{n_2}\right)}\times\text{'t'error df at }5\%\,(1\%)$$

Critical difference between Tr1, Tr3 at 5% (1%)

$$\sqrt{EMS\left(\frac{1}{n_1}+\frac{1}{n_3}\right)}\times\text{'t'error df at }5\%\,(1\%)\text{ and so on}$$

Multiple comparisons

If we reject the null hypothesis in the one-way ANOVA, then we need to establish which group means differ. This will involve conducting a number of tests, but the more tests that we perform, the more likely it is that we will obtain a significant *P*-value on the basis of chance alone.

We have to approach this problem of **multiple comparisons** in such a way that we avoid spurious *P*-values. Formal multiple comparison techniques should be used, such as **Duncan’s multiple range test (DMRT), Least significant difference (LSD) test**, **Bonferroni’s correction**, and **Scheffe’s, Tukey’s** or **Newman–Keuls tests**, often termed procedures.

Two-way analysis of variance Assumptions

A two-way ANOVA test is a statistical analysis tool that determines the effect of two variables on an outcome, as well as testing how altering the variables will affect the outcome.

Assumptions of Two-way ANOVA

When using the two-way ANOVA test, a person must make several assumptions, including:

i. **Independence of variables:** The **two variables for testing** should be independent of each other. One should not affect the other, or else it could result in skewness.

   This means that one cannot use the two-way ANOVA test in settings with categorical variables.

ii. **Homoscedasticity:** In a two-way ANOVA test, the variance should be homogenous. The variation around the mean for each set of data should not vary significantly for all the groups.

iii. **Normal distribution of variables:** The two variables in a two-way ANOVA test should have a normal distribution. When plotted individually, each should have a bell curve. If the data does not meet this criterion, one could attempt statistical data transformation to achieve the desired result.

Summary: differences between one-way and two-way ANOVA

The key differences between one-way and two-way ANOVA are summarized clearly below.

1. A one-way ANOVA is primarily designed to enable the equality testing between three or more means. A two-way ANOVA is designed to assess the **interrelationship of two independent variables on a continuous dependent variable.**

2. A one-way ANOVA only involves **one factor or independent variable**, whereas there are **two independent variables in a two-way** ANOVA.

3. In a one-way ANOVA, **the one factor** or independent variable analysed has three or more categorical groups. **A two-way ANOVA instead compares multiple groups of two factors.**

4. One-way ANOVA need to satisfy only two principles of design of experiments, i.e. replication and randomization. **As opposed to two-way**

ANOVA, which meets all three principles of design of experiments which are replication, randomization and local control, so it is more precise.

The following ANOVA table illustrates the relationship between the sum of squares for each component and the resulting F-statistic for testing the three null and alternative hypotheses for a two-way ANOVA.

1. H_0: There is no interaction between factors

   H_1: There is a significant interaction between factors
2. H_0: There is no effect of Factor A on the response variable

   H_1: There is an effect of Factor A on the response variable
3. H_0: There is no effect of Factor B on the response variable

   H_1: There is an effect of Factor B on the response variable

Source of variation	**df**	**Sum of squares**	**Mean sum of square**	**F**
Factor A	k-1	SSA	$MSA = \frac{SSA}{k-1}$	$F_A = \frac{MSA}{MSE}$
Factor B	*l*-1	SSB	$MSB = \frac{SSB}{l-1}$	$F_B = \frac{MSB}{MSE}$
Interaction AB	(k-1) (l-1)	SSAB	$MSAB = \frac{SSAB}{(k-1)(l-1)}$	$F_{AB} = \frac{MSAB}{MSE}$
Error	kl(m-1)	SSE	$MSE = \frac{SSE}{kl(m-1)}$	
Total	klm -1	SSTo		

Summary of this chapter

- The *F*-test, often called the variance ratio test, may be used to investigate the homoscedasticity of two data sets.
- We determine this cut-off value formally, under the null hypothesis that the two population variances are equal, by referring the ratio to the table of the F-distribution. The degrees of freedom are ($n_1 - 1$) in the numerator (the larger variance) and ($n_2 - 1$) in the denominator (the smaller variance), where n_1 and n_2 represent the two sample sizes.

- A two-way ANOVA test is a statistical analysis tool that determines the effect of two variables on an outcome, as well as testing how altering the variables will affect the outcome.

Problem 1.

Test the equality of variance for the following data:

Sample I : 65 56 45 38 47 68 54 35 78 80

Sample II: 64 30 66 45 55 75 45 25 48 77 65 85

Solution

1. Hypothesis

   H_0: $s_1^2 = s_2^2$

   H_1: $s_1^2 \neq s_2^2$

2. Assumption: a. Sample should be drawn randomly.

   b. Population should be randomly distributed.

3. Level of significance: F vale at significance level at 0.05% is 3.07 at 11, 9 degree of freedom.

4. Calculation:

Sl.no,	**1**	**2**	**3**	**4**	**5**	**6**	**7**	**8**	**9**	**10**	**SUM**		
X	65	56	45	38	47	68	54	35	78	80	566		
X Square	4225	3136	2025	1444	2209	4624	2916	1225	6084	6400	34288		
Variance	= [34288-566*566/10]/9	=250.27											
	1	2	3	4	5	6	7	8	9	10	11	12	SUM
Y	64	30	66	45	55	75	45	25	48	77	65	85	680
Y Square	4096	900	4356	2025	3025	5625	2025	625	2304	5929	4225	7225	42360
Variance	=[42360-680*680/12]11	=347.88											
F value	=347.88/250.27	1.39	at 11 and 9 df										

5. **Conclusion**: Calculated value of F is 1.39 which is less than table value of F, at 0.05 % level is 3.07 at 11, 9 degree of freedom. So H0 is accepted and test is non-significant.

Problem 2:

The average milk production of per day at two farm is recorded as follows:

Farm A: 4 5 8 3 7 6 5 8 4

Farm B: 6 6 7 8 9 7 6 8 7

Test for equality of variances.

Solution

1. Hypothesis:

   H_0: $s_1^2 = s_2^2$

   H_1: $s_1^2 \neq s_2^2$

2. Assumption: a. Sample should be drawn randomly.

   b. Population should be randomly distributed.

3. Level of significance: F vale at significance level at 0.05 % is 3.44 at 8, 8 degree of freedom.

4. **Calculation:**

Sl.No.	**1**	**2**	**3**	**4**	**5**	**6**	**7**	**8**	**9**	**SUM**
X	4	5	8	3	7	6	5	8	4	50
X Square	16	25	64	9	49	36	25	64	16	304
Variance	=[304-50*50/9]/8	=3.279								
	1	2	3	4	5	6	7	8	9	SUM
X	6	6	7	8	9	7	6	8	7	64
X Square	36	36	49	64	81	49	36	64	49	464
Variance	=[464-64*64/9]/8	=1.11								
F Ratio	2.95	at 8 and 8 df								

5. **Conclusion:** Calculated value of F is 2.95 which is less than table value of F, at 0.05% level is 3.44 at 8,8 degree of freedom. So H_0 is accepted and test is non-significant.

13

Design of Experiment

Design of Experiment was first elaborated by R. A. Fisher (1935).

- **Design of Experiment:** Proper allotment of treatments to the available experimental units.
- **Treatment:** It is the factor whose effect is to be studied.

  Example: Application of drugs, feeds, conditions, methods, etc.
- **Experimental units:** The subjects or individuals upon which an experiment is conducted.

Example: Laboratory animals like mice, rat, monkeys, primates; livestock like cattle, buffalo, sheep, goat, poultry; plants, a plot of land, etc.

Necessity of designing an experiment

1. To get maximum information from the available resources.
2. To know the effect of treatment on experimental units and to make comparison between treatments.
3. To know whether the difference between the effect of two treatments is significant or not.
4. A well designed experiment has a well defined method of statistical analysis of data.
5. To minimize error in the whole experimental procedure.
6. An ill designed experiment is not likely to give proper estimates of parameters and comparisons between treatments with desired precision even if the method of data analysis is very good.
7. The design of experiment is essential to get all the important treatment effects to be compared independently.

Important considerations

1. Objectives of the study
2. Types of experimental materials available
3. Characters to be studied
4. Treatments to be undertaken for comparison
5. Precision required

Methods to increase the efficiency of experiment

1. Objective of the experiment, Variability in the experimental units, Precision required and comparison to be made among various treatment effects must be very clear to the experimenter.
2. The design shall be least complicated.
3. When new treatments are under investigation, a tested treatment shall be kept as a control.
4. Statistical analysis of data should be easy as far as possible.
5. Use of auxiliary character, if any.

Principles of designing an experiment

1. **Randomization** in the allotment of treatment to the experimental units.
2. **Replication** under each treatment.
3. By using **local control (also known as blocking), grouping of experimental units** is made to make them homogeneous under each group.

1. Randomization

Randomization means unbiased estimation, or equal chance of selection for each sample, is a process employed to avoid biasness in the allotment of treatments to the experimental units.

Treatment is allotted randomly means with equal chance to each experimental unit. E.g.

Control	F1	F2	F3	F4	Total
Kids	5	5	5	5	20

Methods of randomization:

i) Lottery system

ii) Using the table of random numbers

Randomization is the process of making something random. Randomization is not haphazard; instead, a random process is a sequence of random variables describing a process whose outcomes do not follow a deterministic pattern, but follow an evolution described by probability distributions. For example, a random sample of individuals from a population refers to a sample where **every individual has a known probability of being sampled. Each sample has an equal probability of being chosen**. This would be contrasted with nonprobability sampling where arbitrary individuals are selected.

In various contexts, randomization may involve:

I. generating a random permutation of a sequence (such as when shuffling cards);

II. selecting a random sample of a population (important in statistical sampling);

III. allocating experimental units via random assignment to a treatment or control condition;

IV. generating random numbers (random number generation); or

V. transforming a data stream (such as when using a scrambler in telecommunications).

2. Replication

In statistics, replication is repetition of an experiment or observation in the same or similar conditions. Replication is important because **it adds information about the reliability of the conclusions or estimates to be drawn from the data**. The statistical methods that assess that reliability rely on replication.

For example, if you select **a person from the population** of a city and measure his/her body height and weight, this leaves almost no room for statistical methods. But replication of this observation through **random sampling of, say, 100 more individuals** from the population of the same city, changes the situation radically: now it is possible to **draw some conclusions about weight, height and their interrelation** in the city population at large.

Use of Replication

i) For proper estimation of error component of variance.

ii) To reduce the magnitude of standard error, thereby to make the design of experiment efficient and precise. E.g., SE=SD/√n, when number of observation increases it reduce standard error.

iii) To decide the number of observation (replication) to be recorded under each treatment.

iv) To decide the number treatments to be tested with the available experimental units.

i) **To provide proper estimation of error component of variance**

Only one observation under each treatment cannot provide the estimation of error.

- Each observation is the sum of the:
  - effect of treatment
  - effect of experimental unit
  - effect of uncontrolled extraneous factor (error)

Influence of factors on lactation milk yield of the cow:

Y1 = C1 + F1 + E1

Y2 = C2 + F2 + E2

Where, Y1 & Y2 = LMY of cow 1 & 2 respectively.

C1 & C2 = Genetic effect of cow no. 1 & 2 respectively.

F1 & F2 = effect of Feed 1 & 2 or treatments.

E1 & E2 = Extraneous factor or error.

(Y1 – Y2) = (C1 – C2) + (F1 - F2) + (E1 – E2)

There is no possibility of getting separate estimation of error.

ii) To reduce the magnitude of standard error (SE): Why?

- Effect of two or more treatments are required to be compared.
- Reduction in SE helps in detecting even smaller difference between two treatment means.

$$SE = \frac{\sigma}{\sqrt{n}} \begin{array}{l}\leftarrow \text{Standard deviation} \\ \leftarrow \text{Number of samples}\end{array}$$

- To increase the efficiency of experiment i.e., to make the design of experiment precise.

iii) To decide the number of observations to be recorded under each treatment.

To make the Design of experiment precise, for safer side, the minimum error degrees of freedom (df) should be at least 10.

In Completely randomized Design (CRD), the minimum error df will be decided by the following equation = t(n – 1)

Where

t = no. of treatment

n = no. of replication per treatment.

Suppose 4 treatments/ feeds are to be compared. If error df is 10, then number of replication per treatment will be:

10=4(n-1) = 4n-4 or, 4n=14

Or, n= 14/4=3.5 ≌ 4

That is the number of replication per treatment = 4

Thus total number of experimental units required, N= nxt=4x4=16

iv) To decide the number treatments to be tested with the available experimental units.

In case of CRD, the error df = t(n – 1) that is t(n -1) = 10

Total number of experimental units, N = nxt

Where, t = no. of treatment & n = no. of replication per treatment

If total no. animals available, N = 20

Then, no. treatments will be decided by allotting at least 1 df or 2 animals per treatment.

Hence, N = nxt or, 20 = 2xt or, t = 20/2 = 10

3. Local control or blocking

Grouping of experimental units is done to reduce the variability among the experimental units within the block. **It makes each experimental block homogenous, from heterogeneous sets of experimental units.**

Example of body weight gain per day in grams

		Feeds				
		Feed 1	**Feed 2**	**Feed 3**	**Feed 4**	**Feed 5**
	RIR	20	21	23	18	19
	Leghorn	18	19	20	17	16
Breeds	Cornish	19	20	21	16	18
	Sussex	18	19	20	17	19
	Ancona	17	18	20	16	18

Completely Randomized design (CRD)

CRD is used when **experimental units form a homogenous group**. It follows first two principle of design of experiments and **not follow blocking, the third principle.**

Example:

i) Day old chicks of the same breed.

ii) Very young Laboratory animals – mice, rats, etc.

iii) In agriculture, several plots of land in the same field.

Allotment of treatments to the experimental units:

Allotment takes place completely at random.

Example

Let there are n units and 't ' treatments, N will be multiple of 't'. Then,

N = nt

Each treatment is allotted with n units at random.

Total no. of experimental units required:

- It is decided by error degrees of freedom.
- For precise test of significance in CRD the minimum error df should be at least 10.

Degrees of freedom

DF for treatment = t – 1

Error DF = t(n – 1)

Total DF = (tn – 1) = N – 1

Analysis of Variance (ANOVA)

Source of variation	Degree of freedom
Between treatment	t-1
Error	t(n-1)=N-t
Total	nt-1=N-1

ANOVA

Source of variation	**Degree of freedom (DF)**	**Sum of square (SS)**	**Mean sum of square (SS/DF)**	**F statistics**
Treatments	k-1	SS treatments	MSS treatments	MSS treatments/ MSS error
Error	n-k	SS error	MSS error	
Total	n-1	SS total		

k = the number of different samples

(or treatments)

ni = the size of sample *i*

Ti =the sum of the values in sample *i*

n = the number of values in all samples

$=n_1+n_2+n_3+$……………………..

$\sum x$ = the sum of the values in all samples

$\sum x^2$ = the sum of the squares of values in all samples

For the *F* distribution:

Degrees of freedom for the numerator = *k* -1

Degrees of freedom for the denominator =*n* - *k*

Between-samples sum of squares:

$$SSB = \left(\frac{T_1^2}{n_1} + \frac{T_2^2}{n_2} + \frac{T_3^2}{n_3} + ... \right) - \frac{\left(\sum x\right)^2}{n}$$

Within-samples sum of squares:

$$SSW = \sum x^2 - \left(\frac{T_1^2}{n_1} + \frac{T_2^2}{n_2} + \frac{T_3^2}{n_3} + \ldots \right)$$

Total sum of squares:

$$SST = SSB + SSW = \sum x^2 - \frac{\left(\sum x\right)^2}{n}$$

- Correction factor (CF) is subtracted from SST, SSB to calculate Corrected SST and SSB.

  $$CF = \frac{\left(\sum x\right)^2}{n}$$

- Variance between samples: MSB=SSB/k-1
- Variance within samples: MSW=SSW/n-k
- Test statistic for a one-way ANOVA test:

  F =MSB/MSW

- **Advantage of CRD**
  1. Easy to design
  2. Analysis of data is simple & straight forward.
  3. Statistical analysis does not become complicated if some of the experimental units fail to provide information.

- **Disadvantage of CRD**
  1. The design requires homogeneous set of experimental units.
  2. If experimental units are not homogenous, the error component will be large and this will make the treatment comparison less efficient.

Randomized Block Design (RBD)

- This is an improvement over CRD.
- This is applied when experimental units are heterogeneous.
- All the experimental units are grouped into different homogeneous blocks according to variability.

- Number of experimental units within each block are the multiple of treatments.
- Allotment of treatment is not completely at random like CRD but the allotment is completely at random within the homogeneous block.

- **It follow all principle of design of experiments**
  - Example: To study the effect of certain feed on body weight gain of day old chicks.
  - Chicks are from different breeds say – New Hampshire, Plymouth Rock, Cornish & Sussex.
  - Total number of chicks = 80
  - Number of chicks in each breed = 20
  - Number of male = 10
  - Number of female = 10
  - Number of Block = 4

Degrees of freedom

- In RBD, each treatment is applied to one experimental unit within each block.
- Number of replication of each treatment is equal to number of blocks.
- Number of treatment = t
- Number of Block = b
- Precise test of significance:

The number of blocks and number of treatments will be such that the error degrees of freedom should be at least 10.

Example

- Number of treatment, t = 4
- Number of block = b
- $10 = (b - 1)x(4 - 1) = (b - 1)x3$ or, $10 = 3b - 3$
- or, $3b = 13$ or, $b = 13/3 = 4.2$ or 4
- Total no. of experimental units = 4x4 = 16

ANOVA

Source of variation	Degree of freedom	Sum of square (SS)	Mean sum of square (SS/DF)	F -ratio
Block	n-1	SS block	MSS block	MSS block/ MSS error
Treatments	k-1	SS treatments	MSS treatments	MSS treatments/ MSS error
Error	(n-1)(k-1)	SS error	MSS error	
Total	nk-1			

Sources of variation	Sum of squares (SS)	Degrees of freedom (d.f)	Mean sum of square (MS)	F-ratio
Between columns	$\sum \frac{(Tj^2)}{Nj} - \frac{(T^2)}{n}$	(c-1)	$\frac{\text{SS between coloumns}}{(c - 1)}$	$\frac{\textit{MS bewteen columns}}{\textit{MS residual}}$
Between rows	$\sum \frac{(Ti^2)}{Ni} - \frac{(T^2)}{n}$	(r-1)	$\frac{\textit{SS betweem rows}}{(r - 1)}$	$\frac{\textit{MS betweem rows}}{\textit{MS residual}}$
Residual error	Total SS-(SS between columns and SS between rows)	(c-1)(r-1)	$\frac{\text{SS residual}}{(c - 1)(r - 1)}$	
Total	$\sum Xij^2 - \frac{(T^2)}{n}$	(c.r -1)		

Merit of RBD

- Most frequently used design. It is easy to design with local control.
- Statistical analysis of data is simple but little bit tricky than CRD.
- It eliminates one assignable causes of variation among experimental units by using local control or blocking.
- **The error DF in RBD is lesser than that of CRD due to removal of DF for blocks**.

Demerit of RBD

- When the **number of treatments to be compared is large** then it becomes difficult to get large blocks of homogeneous experimental units.

- The analysis of data becomes difficult if one or more than **one observation in RBD is missing.**
- Under such condition it will be necessary to remove or omit that block.

Example 1: One way classification of ANOVA (CRD)

Milk yield of three groups of cattle

Sample group	Gir	Sahiwal	Kankrej
	3	9	1
	7	12	2
	7	11	6
	6	8	4
	2	5	7
Total	25	45	20
Mean	5	9	4
Observation(n)	5	5	5

Calculation

1. Null hypothesis H_0, there is no difference in milk yield among three breeds of cattle. H_0: $\overline{X}1=\overline{X}2=\overline{X}3$

   Alternate hypothesis H1, three sample means are not equal

   H_1: $\overline{X}1\neq\overline{X}2\neq\overline{X}3$

2. Significance level 5% is used, Critical F value is 3.88 at 2 and 12 degree of freedom.

3. Assumptions:

   All samples are drawn randomly.

   Parent populations are normally distributed.

   Various treatments and environmental effects are additive in nature.

4. Calculations:

   Correction Factor (CF)= (90x90)/15=540

   TSS=3*3+7*7+7*7+6*6+2*2+9*9+12*12+11*11+8*8+5*5+1*1+2*2 +6*6+4*4+7*7

   =9+49+49+36+4+81+144+121+64+25+1+4+36+16+49

   =688

   Corrected TSS=688-540=148 (**CF subtracted**)

BSS=25*25/5+45*45/5+20*20/5

=125+405+80=610

Corrected BSS=610-540=70 (**CF subtracted**)

WSS=148-70=78(CF subtracted)

ANOVA table

Source of variation	Degree of freedom	Sum of square	Mean sum of square	F
Between	2	BSS=$\sum(Y^2i)/ni - \sum(Y)^2/N$ =70	BSS/DF=35	F=35/6.5=5.38
Within	12	WSS=$\sum(Y^2) - \sum(Y)^2/N$ =78	WSS/DF=6.5	
Total	14	TSS=148		

5. **Conclusion:** The "F" value is inside the critical region and H_0 is rejected, alternate hypothesis is accepted, explain that three breeds have significant differences in milk yield.

Example 2: For two way classification (The randomized block design) the ANOVA will be

Source of variation	Degree of freedom	Sum of square	Mean sum of square (SS/DF)	F
Block	n-1	SS block	MSS block	MSS block/ MSS error
Treatments	k-1	SS treatments	MSS treatments	MSS treatments/ MSS error
Error	(n-1)(k-1)	SS error	MSS error	
Total	nk-1			

Problem: Twenty experimental pigs allocated to four farms (five at each) and five different rations were assigned at random to the pig at all the farms. The gain in weights were recorded as follows:

	Rations	Rations	Rations	Rations	Rations	
Farms	1	2	3	4	5	total
Farm A	167	166	138	201	193	865
Farm B	170	170	153	165	164	822
Farm C	130	171	160	187	200	848
Farm D	151	169	172	166	142	800
Total	618	676	623	719	699	3335

Calculation

1. Null hypothesis H_0, there is no difference in effect of five ration among four farms of pigs. $H_0: \bar{X}_1 = \bar{X}_2 = \bar{X}_3 = \bar{X}_4 = \bar{X}_5$ also there is no differences among four farms.

Alternate hypothesis H_1, five sample means among four farms are not equal $H_1 : \bar{X}_1 \neq \bar{X}_2 \neq \bar{X}_3 \neq \bar{X}_4 \neq \bar{X}_5$. There is differences among four farms.

2. Significance level 5% is used, Critical F value is degree of freedom.
3. Assumptions:

   All samples are drawn randomly.

   Parent populations are normally distributed.

   Various treatments and environmental effects are additive in nature.
4. Calculations:

Ration total (XR1)	R1	R2	R3	R4	R5
	618	676	623	719	699
Farm total (XT1)	T1	T2	T3	T4	
	865	822	848	800	

Grand Total ($\sum Y$) =3335 Crude total sum of square ($\sum X^2$)=562845

Correction factor (CF)=$(3335)^2/20$=556111.25

Ration SS=$618^2/4+ 676^2/4+ 623^2/4 +719^2/4 + 699^2/4$ -CF

=558147.75-556111.25

=2036.5

Farm SS=$865^2/5+ 822^2/5+ 848^2/5+ 800^2/5$ -CF

=556602.6-556111.25

=491.35

Corrected Total SS= TSS - CF= 562845-556111.25

=6733.75

Within SS= Total SS – Ration SS – Farm SS=6733.75- 2036.5 - 491.35

=4205.9

ANOVA Table

Source of variation	Degree of freedom	Sum of square	Mean sum of square	F
Ration	5-1=4	2036.5	509.1	1.45 (non significant)
Farm	4-1=3	491.35	63.78	0.46 (non significant)
Error	(5-1)(4-1)=12	4205.9	350.5	
Total	20-1=19	6733.75		

Conclusion: The table value of F at 4 and 12 degree of freedom is 3.26 is more than the calculated value. The effect of ration on weight gain is non-significant.

The table value of F at 3 and 12 degree of freedom is 3.49 is more than the calculated value. The effect of farm on weight gain is non-significant.

Latin square design

Basic features

1. It is improvement over RBD by eliminating more source of variation.
2. Elimination of two sources of vitiations.
3. If k treatments replicated k times, the total number of experimental unit is k^2.

The rules for the Latin square design are as follows

I. Randomly assigned treatments within rows and columns, and each treatment once per row and once per column.

II. Equal numbers of rows, columns, and treatments.

III. Suitable for controlling variation in two different directions.

What is the use of Latin square?

Latin square design works by blocking the experimental units into rows and columns, and assigning the levels of the two factors randomly within each block. This way, the variation due to the row and column factors is accounted for and removed from the error term.

A 3 x 3 Latin square is a grid of size 3 x 3, that is filled with three different symbols in such a way that no row has duplicate symbols and no column has duplicate symbols, in other words each row and each column containing all three symbols.

Why is it called a Latin square?

The Latin Square Design gets its name from the fact that we can write it as a square with Latin letters to correspond to the treatments. The treatment factor levels are the Latin letters in the Latin square design. The number of rows and columns has to correspond to the number of treatment levels.

The name "Latin square" was inspired by mathematical papers by **Leonhard Euler** (1707–1783), who used Latin characters as symbols, but any set of symbols can be used: in the above example, the alphabetic sequence A, B, C can be replaced by the integer sequence 1, 2, 3. Euler began the general theory of Latin squares.

ANOVA of k x k latin sqare design

Source of variation	**Degree of freedom**	**Sum of square**	**Mean sum of square**	**F**
Rows	k-1	$\sum R^2i/k$-CF	SSR/K-1	
Column	k-1	$\sum C^2i/k$-CF	SSC/K-1	
Treatments	k-1	$\sum T^2i/k$-CF	SST/K-1	MSST/MSSE
Error	(k-1)(k-2)	By substruction from total	SSE/(k-1)(k-2)	
Total	K^2-1	$\sum Y^2ij/k$-CF		

Example: Petrol consumption of different makes of Cars have been illustrated with five makes of cars. Di (i=1 2, 3, 4 to D5) denotes drivers driving on i th day. A, B, C, D and E denotes the makes of the car.

Speeds in mile per hour						
		25	35	50	60	70
Drivers and Days	D1	B 19.5	E 21.7	A 18.1	D 14.8	C 13.7
	D2	D 16.2	B 19.0	C 16.3	A 17.9	E 17.5
	D3	A 20.6	D 16.5	E 19.5	C 15.2	B 14.1
	D4	E 22.5	C 18.5	D 15.7	B 16.7	A 16.0
	D5	C 20.5	A 19.5	B 15.6	E 18.7	D 12.7

SS due to Speeds =7719.49-7638.76=80.73

SS due to Drivers =7640.12-7638.76=1.36

SS due to Makes =7704.18-7638.76=65.42

Total SS =7792.70-7638.76=153.94

Error SS =153.94 - 80.73 - 1.36 - 65.42 = 6.43

ANOVA of 5 x 5 latin sqare design

Source of variation	**Degree of freedom**	**Sum of square**	**Mean sum of square**	**F**
Rows	4	80.73	20.18	
Column	4	1.36	0.34	
Treatments	4	65.42	16.35	30.28**
Error	12	6.42	0.54	
Total	24	153.94		

** highly significant.

Mean number of miles per gallon for different makes arranged in under:

E	A	B	C	D
19.98	18.42	16.98	16.84	15.18

SE= √ 0.54/5=0.33

CD at 1 per cent =3.055 √ 2x0.54/5=1.42, where 3.055 is table value of 't' for error d.f. 12 at 1% level.

Initial differences indicate that car make E is significantly better that other makes. Car make A is better than B, C, D makes and D makes is worst of all.

Critical Difference (C.D.) test

If F is significant or highly significant, critical difference between treatment means is to be worked out. Critical difference between any two treatment means is defined as the least significant difference between any two treatment means, to be exceeded by the difference between two treatment means to declare them as significantly different.

Critical difference between any two treatment means at 5% level = Standard error of the difference between the treatment means x table value of 't' for error d.f. at 5% level.

Critical difference between any two treatment means at 1% level = Standard error of the difference between the treatment means x table value of 't' for error d.f. at 1% level.

Critical difference between Tr1 and Tr2 at 5% (1%)

$$\sqrt{EMS\left(\frac{1}{n_1}+\frac{1}{n_2}\right)} \times 't' \, error\, df\ at\, 5\%\, (1\%)$$

Summary

A well-designed experiment likely to give proper estimates of parameters and comparisons between treatments with desired precision.

The design of experiment is essential to get all the important treatment effects to be compared independently.

Partitioning of variance into different components in ANOVA, help to determine error component, effects of rows, column, and treatments for F test.

CRD is used when experimental units form a homogenous group is just like one way ANOVA.

Randomized Block Design (RBD), this is an improvement over CRD. This is applied when experimental units are heterogeneous. All the experimental units are grouped into different homogeneous blocks according to variability.

Latin square design (LSD), It is improvement over RBD by eliminating more source of variation. If k treatments replicated k times, the total number of experimental unit is k^2. Randomly assigned treatments within rows and columns, and each treatment once per row and once per column. Suitable for controlling variation in two different directions.

Difference between CRD and RBD

		CRD	**RBD**
1	Layouts	The whole field is divided directly into plots (product of replications=treatments).	The field is divided into homogeneous i.e, plots=number of treatments.
2	Randomization	In CRD, treatment-wise.	In RBD, replication or block-wise.
3	Local Control (Blocking)	Not adopted in this case.	In this case, adopted by homogeneous blocks.
4	Components of Variation	Divided into 2 component divisions. Treatment and Error	Divided into 3 component divisions. Row, column and error
5	Analysis	Analysis in CRD is very easy and simple.	In this case, analysis is easy but becomes complicated if the missing plot technique is applied.
6	Accuracy	CRD is less accurate than other designs.	The accuracy level of RBD is greater than CRD.
7	Efficiency	CRD is not that efficient when compared to RBD.	RBD is very efficient than CRD.
8.	ANOVA	One way ANOVA	Two way ANOVA

14

Sampling and Livestock Census

Important Definitions

• **Sampling method** or **sampling technique** is the process of studying the population by gathering information (on basis of samples) and analyzing that data.

- Population: The whole group we are interested in census.
- Census: A collection of data from the **whole** population
- Sample: A collection of data from **part** of the population
- Parameter Vs Statistics: Parameter is statistical measure pertaining to population whereas statistics is measure pertaining to sample.
- Estimate-Value of population parameter obtained from sample.
- Livestock census carried every 5 years (2019-20th livestock census)

20th livestock Census (2019), last census conducted in India

- For the first time, livestock data were collected on line in 20th Livestock Census.
- It was designed to capture Breed-wise number of animals and poultry birds.
- Dog, Rabbit and elephant were included for first time.
- The total Livestock population is 535.78 million
- The total number of cattle in the country is 192.49 million(^) (West Bengal)
- The total buffaloes in the country is 109.85 Million (^) (Uttar Pradesh)
- The total sheep in the country is 74.26 Million (^) (Telangana)
- The Goat population in the country in 2019 is 148.88 Million (^) (Rajasthan)

- The total Pigs in the country is 9.06 Million(v)
- The total Horses and Ponies in the country is 3.4 Lakhs (v)

((^) means increasing, (v)means decreasing)

- The Female Cattle (Cows population) is 145.12 million, increased by 18.0% over the previous census (2012).
- The Exotic/Crossbred and Indigenous/Non-descript Cattle population in the country is 50.42 million and 142.11 million respectively.
- The Indigenous/Non-descript female cattle population has increased by 10% in 2019 as compared to previous census.
- The population of the total Exotic/Crossbred Cattle has increased by 26.9 % in 2019 as compared to previous census.
- There is a decline of 6 % in the total Indigenous (both descript and non-descript) Cattle population over the previous census. However, the pace of decline of Indigenous Cattle population during 2012-2019 is much lesser as compared to 2007-12 which was about 9%.
- The total milch animals (in-milk and dry) in cows and buffaloes is 125.34 Million, an increase of 6.0 % over the previous census.
- The total sheep in the country is 74.26 Million in 2019, increased by 14.1% over previous Census.
- The total Mithun in the country is 3.9 Lakhs in 2019, increased by 30.0% over previous Census.
- The total Yak in the country is Fifty Eight Thousand in 2019, decreased by 24.67% over previous Census.
- The total population of Mules in the country is Eighty Four Thousand in 2019, decreased by 57.1% over previous Census.
- The total population of Donkeys in the country is 1.2 Lakhs in 2019, decreased by 61.23% over previous Census.
- The total Camel population in the country is 2.5 Lakhs in 2019, decreased by 37.1% over previous Census.
- The total Poultry in the country is 851.81 Million in 2019, increased by 16.8% over previous Census.
- The total Backyard Poultry in the country is 317.07 Million in 2019, increased by 45.8% over previous Census.

- The total Commercial Poultry in the country is 534.74 Million in 2019, increased by 4.5% over previous Census.

Table 1: Livestock and Poultry population, 2012 and 2019 of major species

Sl. No.	Species	19thLivestock Census 2012 (no. in millions)	20thLivestock Census 2019 (no. in millions)	Growth Rate (%) 2012-19
1	Cattle	190.90	193.46	1.34
2	Buffalo	108.70	109.85	1.06
3	Yaks	0.08	0.06	-24.90
4	Mithun	0.30	0.39	29.52
	Total Bovines	**299.98**	**303.76**	**1.26**
5	Sheep	65.07	74.26	14.13
6	Goat	135.17	148.88	10.14
7	Pigs	10.29	9.06	-12.03
8	Horse & Ponies	0.63	0.34	-45.58
9	Mule	0.20	0.08	-57.09
10	Donkey	0.32	0.12	-61.23
11	Camel	0.40	0.25	-.37.05
	Total livestock	**512.06**	**535.78**	**4.63**
12	Poultry	729.21	851.81	16.81

Table 2: Livestock population, 2012 and 2019 of major states

Sl. No.	State	Population (no. in millions) 2012	Population (no. in millions) 2019	% Change
1	Uttar Pradesh	68.7	67.8	-1.35
2	Rajasthan	57.7	56.8	-1.66
3	Madhya Pradesh	36.3	40.6	11.81
4	West Bengal	30.3	37.4	23.32
5	Bihar	32.9	36.5	10.67
6	Andhra Pradesh	29.4	34.0	15.79
7	Maharastra	32.5	33.0	1.61
8	Telengana	26.7	32.6	22.21
9	Karnataka	27.7	29.0	4.70
10	Gujarat	27.1	26.9	-0.95

Advantage of sampling

1) Save time
2) Less costly
3) Provides result with good accuracy
4) When population is infinite, articles to be destroyed for testing quality, census method cannot be used and hence sampling need to be done.

Disadvantage

1) We only get estimate of population parameter from sampling which may differ from actual value.
2) Sampling Error: A sampling error occurs when the sample used in the study is not representative of the whole population.
3) The standard error (SE) of a statistic is the approximate standard deviation of a statistical sample population. ($SE = \sigma / \sqrt{n}$ where σ = the population standard deviation $\sqrt{n}$ = the square root of the sample size

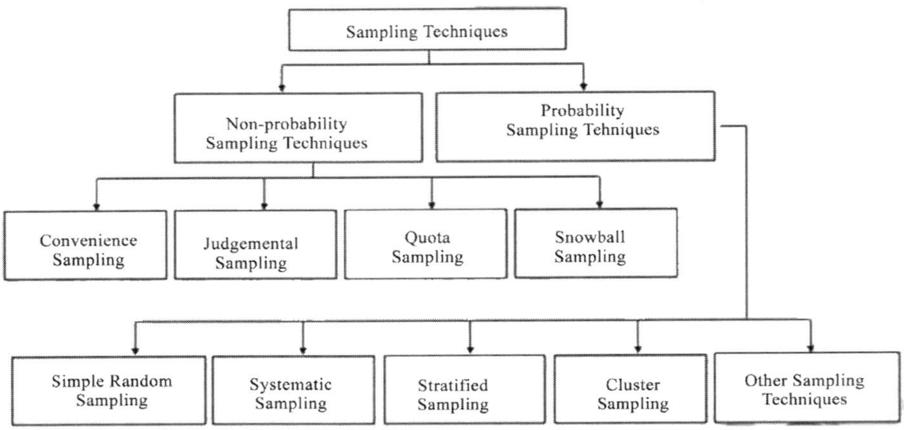

Probability Sampling

I. In this method, all the eligible individuals have a chance of selecting the sample from the whole sample space.
II. This method is more time consuming and expensive than the non-probability sampling method.
III. The benefit of using probability sampling is that it guarantees the sample that should be the representative of the population.

Simple Random Sampling

I. Every item in the population has an equal and likely chance of being selected in the sample.
II. Since the item selection entirely depends on the chance, this method is known as "**Method of chance Selection**".
III. As the sample size is large, and the item is chosen randomly, it is known as "**Representative Sampling**".

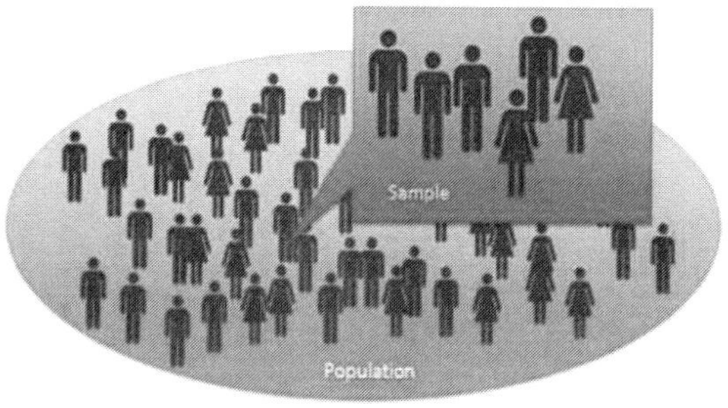

Systematic Sampling

I. In the systematic sampling method, the items are selected from the target population by selecting the random selection point and selecting the other methods after a fixed sample interval.

II. It is calculated by dividing the total population size by the desired population size.

Stratified Sampling

I. In a stratified sampling method, the total population is divided into smaller groups to complete the sampling process.

II. The small group is formed based on a few characteristics in the population. After separating the population into a smaller group, the statisticians randomly select the sample.

Clustered Sampling

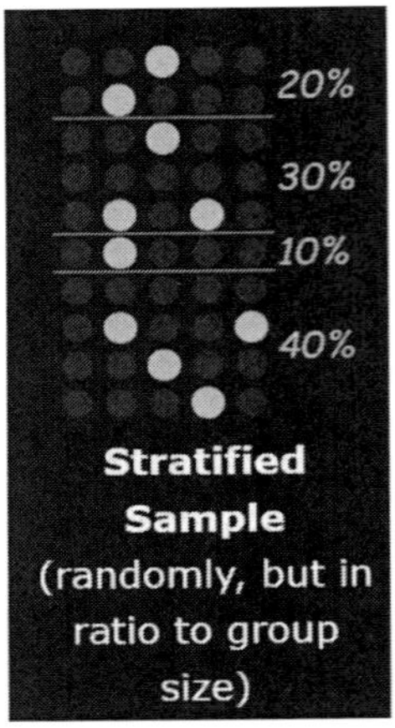

I. In the clustered sampling method, the cluster or group of people are formed from the population set.

II. The group has similar significatory characteristics. Also, they have an equal chance of being a part of the sample.

III. This method uses simple random sampling for the cluster of population.

(We break the population into many groups, then **randomly choose whole groups**)

Non-Probability Sampling

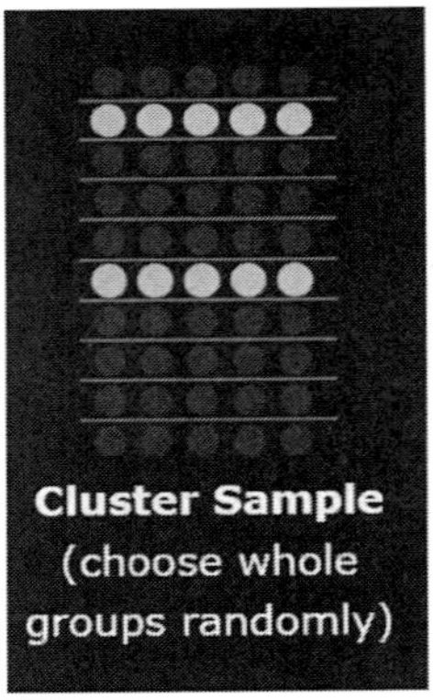

I. The non-probability sampling method is a technique in which the researcher selects the sample based on subjective judgment rather than the random selection.

II. In this method, not all the members of the population have a chance to participate in the study

Convenience Sampling

I. In a convenience sampling method, the samples are selected from the population directly because they are conveniently available for the researcher.

II. The samples are easy to select, and the researcher did not choose the sample that outlines the entire population.

Purposive or Judgmental Sampling

I. In purposive sampling, the samples are selected only **based on the researcher's knowledge.**

II. As their knowledge is instrumental in creating the samples, there are the chances of obtaining highly accurate answers with a minimum marginal error.

III. It is also known as judgmental sampling or authoritative sampling.

Quota Sampling

I. In the quota sampling method, the researcher forms a sample that involves the individuals to represent the population based on specific traits or qualities.

II. The researcher chooses the sample subsets that bring the useful collection of data that generalizes the entire population.

Snowball Sampling

I. Snowball sampling is also known as a chain-referral sampling technique.

II. In this method, the samples have traits that are difficult to find. So, each identified member of a population is asked to find the other sampling units.

III. Those sampling units also belong to the same targeted population.

Livestock Census : Background

Coverage

Livestock census covers the entire nation in almost 648 districts and nearly 660 thousand villages and 64 thousand urban wards across the country Census conducted with the following schedules

Schedule Description

1 House-list of households, enterprises & institutions Recocing list of households, etc. under enumeration.

2 Village/Ward profiles Based on schedule 1 & schedule 3

3 Household, Enterprise & Institution schedule Further divided into three parts

   i. Livestock

   ii. Poultry

   iii. Number of Equipment Used

Livestock Census 2012: Objectives and targets

- To provide information on livestock population, category wise along with age, sex-composition etc. up to village level in rural areas and town level in urban areas.

- Animals and Poultry birds owned by the households, household enterprises, non-household enterprises and institutions were counted at their site.
- 100% scrutiny of the schedules.

19th Livestock Census: Implementation (2012)

- 100% Central assistance to States/UTs
- About 171 thousand enumerators and 28472 supervisors were engaged for LC.
- For Breed Survey Para-vets/ Veterinary services practitioners were engaged for Enumeration
- State Nodal Officers- Nodal officer was designated as the Livestock Census Authority, who was the overall in-charge of supervising the conduct of the Census and Breed Survey, its scrutiny and validation.
- NIC: for software development, Data entry and report generation.

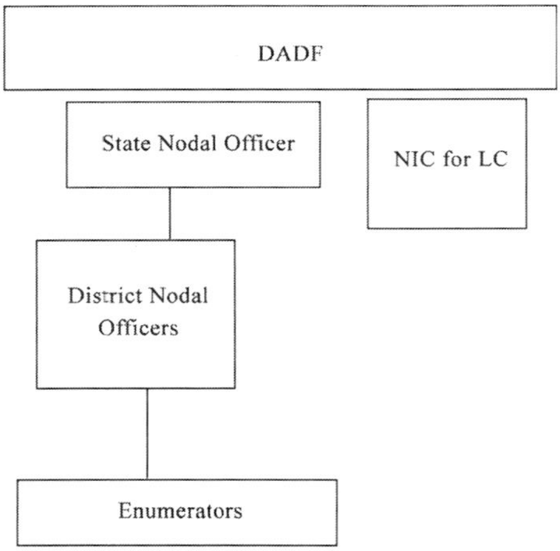

Integrated Sample Survey (ISS): Overview

- Livestock products data is estimated through "Integrated Sample Survey"
- ISS helps State Animal Husbandry Departments to assess the impact of the government efforts towards
- Increase production.

- Genetic up gradation.
- Disease Control
- Feed & fodder management
- ISS helps in predicting GDP from this sector

Methodology of Integrated Sample Survey (ISS) : Coverage

- ISS covers entire Rural and urban areas
- Conducted in sample villages/urban wards.
- Undertaken from March to February, divided into 3 seasons
- Gives progressive estimates while taking care of season-variability
- Each season is further divided into 4 rounds (or months).

Season Period Duration

Summer March 1 – June 30 122 days

Rainy July 1 – October 31 123 days

Winter November 1 – February 28 (29) 120 (121) days

Methodology of Integrated Sample Survey (ISS) : Design

• **Stratified 3 stage design with district as stratum**

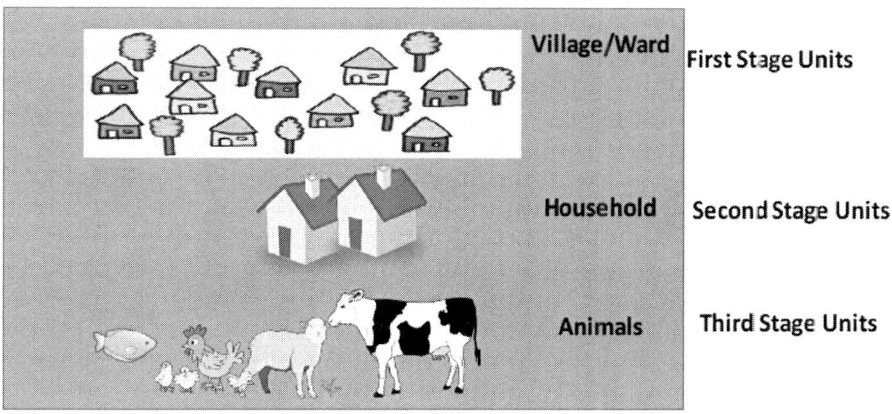

ISS : Sample Selection of First Stage Units

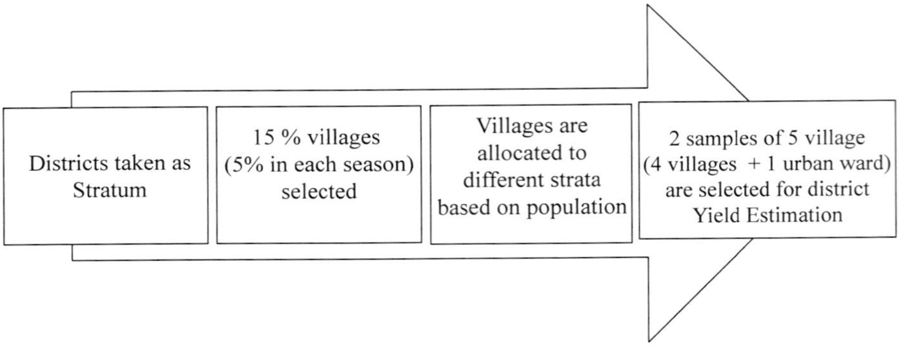

ISS : Sample Selection for Second Stage Units

- Sample Selection for Second units are done with equal probability and without replacement for estimation of Yield rate.
- The recording of wool yield will be done in the shearing season in the selected villages from the sample of 5/8 households having sheep.

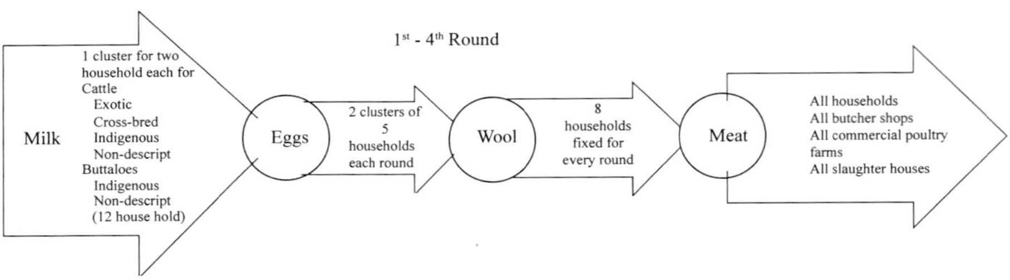

20th Livestock Census (2019)

- The Department of Animal Husbandry & Dairying under Ministry of Fisheries, Animal Husbandry & Dairying attributes critical importance to livestock and to the collection and availability of up-to-date and accurate data
- The livestock census is conducted across the **country periodically since 1919.**
- **So far, 20 Livestock Censuses were conducted** in participation with State Governments and UT Administrations.
- The 20th Livestock Census was launched during the month of October, 2018 and completed in 2019.

- Livestock census carried every 5 years (2019-20th livestock census)
- Another important feature of 20th Livestock Census is it has been designed to capture Breed-wise number of animals and poultry birds.
- For the first time, livestock data were collected on line in 20th Livestock Census.
- Advance technology has been adopted to collect data through tablet computers.
- The 20th Livestock Census was carried out in about 6.6 lakhs villages and 89 thousand urban wards across the country covering more than 27 Crores of Households and Non-Households.

What are the Key Highlights?

Milk Production

- Total milk production in India was 221.06 million tons in 2021-2022, keeping It the largest milk producing country in the world.
- Production had increased by 5.29% over the previous year.
- The Indigenous cattle contribute 10.35% of the total milk production in the country whereas non-descript cattle contribute 9.82% and non-descript buffaloes contribute 13.49% of the total milk production in the country.
- In India, the milk production breakdown for different animals in 2024 is as follows:
- Cows: Contribute approximately 47.3% of the total milk production.
- Buffaloes: Account for around 49.0% of the milk production (major contributor).
- Goats: Contribute a smaller share, approximately 3.0%, to the overall milk production
- Top five major milk producing States are Rajasthan (15.05%), Uttar Pradesh (14.93%), Madhya Pradesh (8.06%), Gujarat (7.56%) and Andhra Pradesh (6.97%).
- India is the largest producer of Milk and Buffalo Meat, the 2nd largest producer of Goat meat, 3rd in Egg production and the 8th largest in overall Meat Production in the world.

- The **total egg** production was 129.60 billion numbers, and it is an increase by 6.19% than the previous year.
- The **total meat** production in the country was 9.29 million tonnes, increasing by 5.62% as compared to the previous year.
- The **total wool** production in the country during 2021-22 was 33.13 thousand tonnes which had declined by 10.30% as compared to previous year.

15

Bioassays

Biological assay or bioassay

- A bioassay is an analytical method to determine the concentration or potency of a substance by its effect on living animals or plants (in vivo), or on living cells or tissues (in vitro).
- Bio-assays are thus a type of experiments with the object of comparing the efficacy of two or more substances, or preparations, like drugs, by using responses produced by them on suitable living organisms.
- The bio assay involves a stimulus applied to a subject and the response of the subject to the stimulus.
- When a stimulus is applied to a subject there may be a change in some characteristics of the subject. Such changes in the subject are known as responses.
- The response may be quantitative as in the case of weight or qualitative as in the case of mortality. The magnitude of response depends upon the dose.
- Normally, two preparations having a common effect are taken for assaying. One of the preparations is of known strength and is called the standard preparation and the other is of unknown strength and is called test preparation.
- The objective of the assay is to estimate the potency of the test preparation relative to that of the standard preparation.

  R: Represents the relative potency.

  R= Zs/ Zt (Zs Dose of Standard, Zt Dose of test)

Analytical vs comparative dilution assay

- An assay with two preparations containing the **same effective ingredient**, which is **responsible for the response**, is called **analytical dilution assay**.

Example: If you're assessing the potency of a drug or a vitamin, an ADA would involve comparing it to a standard sample with a known concentration of the active ingredient.

- An assay with two preparations which have a common effect but **do not contain the same effective ingredient**, is called a **comparative dilution assay**.

  Example: Suppose you're comparing the efficacy of two different fungicides. While they may have different active ingredients, their overall impact on fungal growth is being assessed.

Types of Bioassay

1. Direct assays
2. Indirect assays
   a) Indirect assays based upon quantitative responses and
   b) Indirect assays based upon quantal responses

(If the measured response is binary, the assay is quantal, if not, it is quantitative)

1. Direct Assay

In a direct assay, the stimulus applied to the subject is specific and directly measurable, and the response to that stimulus is recorded.

The variable of interest is the specific stimulus required to produce a response of interest (ex. death of the subject).

2. Indirect assay

In an indirect assay, the stimulus is fixed in advance and the response is measured in the subjects.

The variable of interest in the experiment is the response to a fixed stimulus of interest.

a) Quantitative response: The measurement of the response to the stimulus is on a continuous scale.

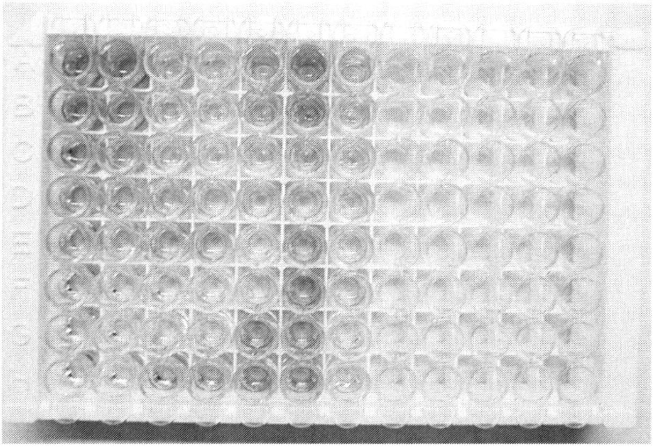

Fig. 1: Showing ELISA plate reading for indirect bioassay

b) Quantal response: The response is binary; it is a determination of whether or not an event occurs.

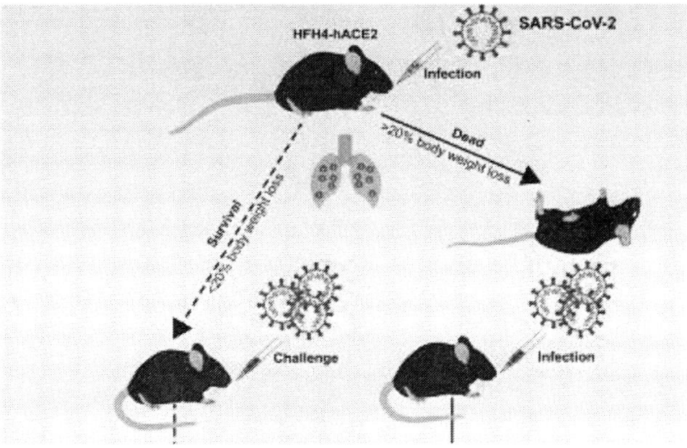

Fig. 2: Showing Quantal response of bioassays live versus death

Some important terms

LD50 LD stands for "Lethal Dose". LD 50 is the amount of a material, given all at once, which causes the death of 50% (one half) of a group of test animals. The LD 50 is one way to measure the short term poisoning potential (acute toxicity) of a material.

Graphical methods (Miller and Tainter)

Step three:

- The probit values are plotted against log-doses, then dose corresponding to probit 5, i.e., 50% is found out
- In the present case the log LD50 is 1.76 and LD50 = 57.54 mg/kg

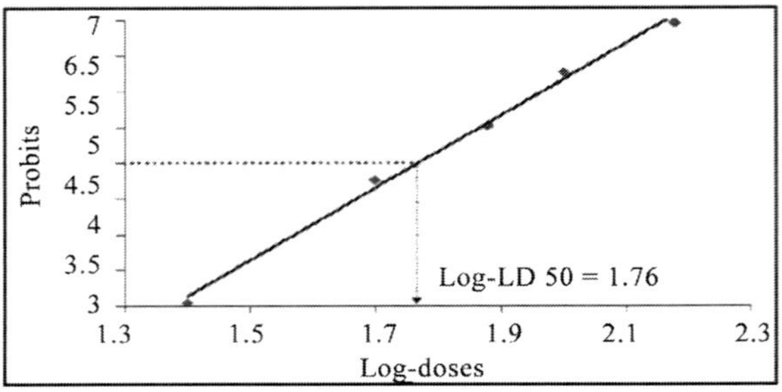

Fig. 3: Calculation of LD50 present Plot of log-doses versus probits

LC50 LC stands for "Lethal Concentration" The concentrations of the chemical in air that kills 50% of the test animals during the observation period (traditional 4 hours) is the LC 50 value.

ED50 ED stands for "Effective Dose". The ED50 is the dose of a medication that produces a desired pharmacologic effect in 50% of the studied patient population that takes the medication.

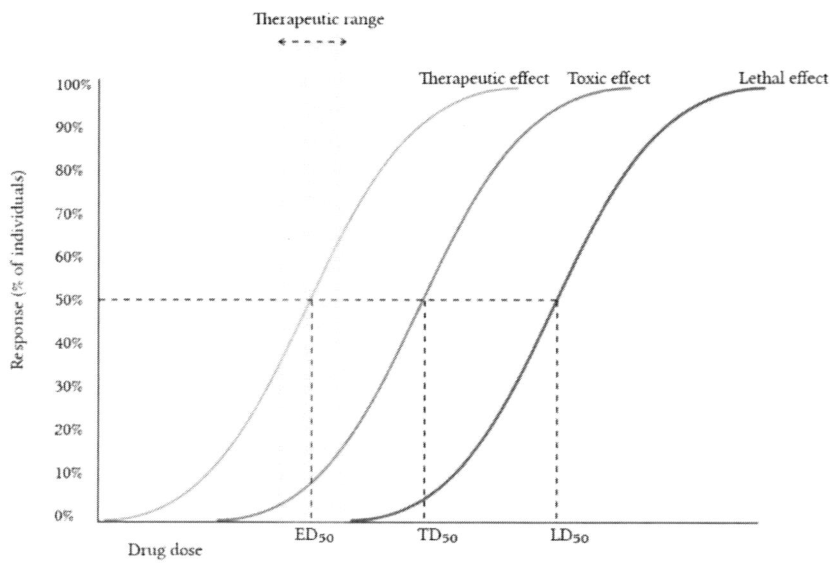

Fig. 4: Calculation ED50, TD50 and LD50 from the graph.

Safety-based therapeutic index

LD50 is ensured to be as high as possible to decrease risk of lethal effects (often leading to death), and for larger therapeutic window. Hence, a higher safety-based therapeutic index indicates a larger therapeutic window.

$$\bullet TI_{safety} = \frac{LD_{50}}{ED_{50}}$$

Efficacy-based therapeutic index: ED_{50} is ensured to be as low as possible for faster drug response and larger therapeutic window, whereas TD_{50} is ensured to be as high as possible to decrease risk of toxic effects of drugs.

$$\bullet TI_{\text{efficacy}} = \frac{ED_{50}}{TD_{50}}$$

Testing of hypothesis

Statistical hypothesis testing is making a decision between rejecting or not rejecting a null hypothesis, on the basis of a set of observations.

Forming of Hypothesis

a. Null Hypothesis (H_0) H-nought

b. Alternative Hypothesis (H_1) or (Ha)

- A **null hypothesis** (typically that there is **no effect**) is compared with an **alternative hypothesis**(typically that there is **an effect**, or that there is an effect of a particular sign).
- For example, in evaluating whether a new treatment works for a disease, the null hypothesis typically would be that the remedy does not work, while the alternative hypothesis would be that the remedy does work.

Types of error

	H_0 Reject	H_0 Accept
H_0 true	Type I error	Correct decision
H_0 false	Correct decision	Type II error

Alpha (α) : Probability of Type-I error

Beta (β) : Probability of Type-II error

Power of test : 1 –β

P value

- If P value < 0.05, it suggests that the test is Significant.

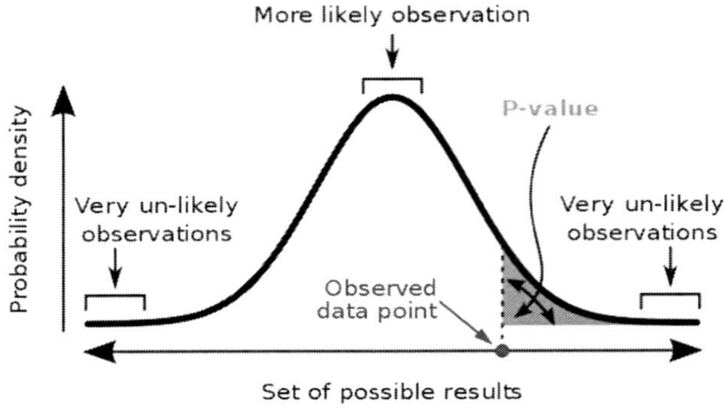

Fig. 5: A p-value (shaded green area) is the probability of an observed (or more extreme) result assuming that a null hypothesis is true.

The p-value tells us how likely it is to observe the test statistic (or a more extreme value) if the null hypothesis were true. A large p-value (greater than 0.05) indicates weak evidence against the null hypothesis. We do not have enough evidence to reject it.

16

Least Square Method and Linear Models

What is Least Square Method?

The Least Squares Method is used to derive a generalized linear equation between two variables, one of which is independent and the other dependent on the former. The value of the independent variable is represented as the x-coordinate and that of the dependent variable is represented as the y-coordinate in a two-dimensional cartesian coordinate system. Initially, known values are marked on a plot. The plot obtained at this point is called a scatter plot. Then, we try to represent all the marked points as a straight line or a linear equation. The equation of such a line is obtained with the help of the least squares method. This is done to get the value of the dependent variable for an independent variable for which the value was initially unknown. This helps us to fill in the missing points in a data table or forecast the data. The method is discussed in detail as follows. The least-squares method was officially discovered and published by Adrien-Marie Legendre (1805), though it is usually also co-credited to Carl Friedrich Gauss (1809), who contributed significant theoretical advances to the method.

Least Square Method Definition

The least-squares method can be defined as a statistical method that is used to **find the equation of the line of best fit related to the given data. This method is called so as it aims at reducing the sum of squares of deviations (residual part) as much as possible**. The line obtained from such a method is called a **regression line**.

The least-squares method is a crucial statistical method that is practised to find a regression line or a best-fit line for the given pattern. This method is described by an equation with specific parameters. The method of least squares is generously used in evaluation of regression equations. In regression analysis, this method is said to be a standard approach for the approximation of sets of equations having more equations than the number of unknowns.

The method of least squares actually defines the solution for the minimization of the sum of squares of deviations or the errors in the result of each equation. Find the formula for sum of squares of errors, which help to find the variation in observed data.

The least-squares method is often applied in data fitting. The best fit result is assumed to reduce the sum of squared errors or residuals which are stated to be the differences between the observed or experimental value and corresponding fitted value given in the model.

There are two basic categories of least-squares problems:

- Ordinary or linear least squares
- Nonlinear least squares

These depend upon linearity or nonlinearity of the residuals. The linear problems are often seen in regression analysis in statistics. On the other hand, the non-linear problems are generally used in the iterative method of refinement in which the model is approximated to the linear one with each iteration.

Least Square Method Graph

In linear regression, the line of best fit is a straight line as shown in the following diagram:

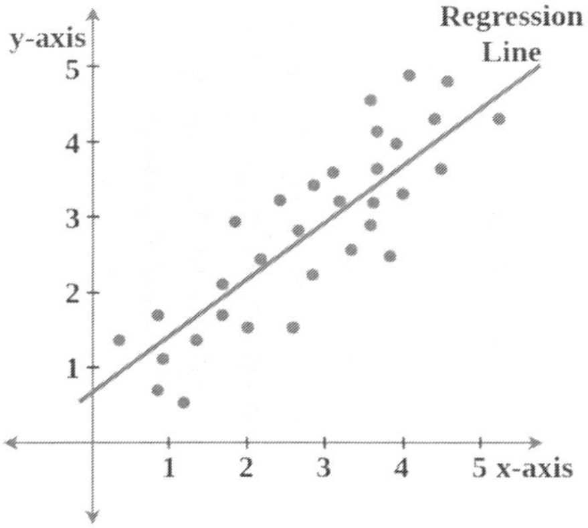

The red points in the above plot represent the data points for the sample data available. Independent variables are plotted as x-coordinates and dependent ones are plotted as y-coordinates. The equation of the line of best fit obtained from the least squares method is plotted as the red line in the graph.

We can conclude from the above graph that how the least squares method helps us to find a line that best fits the given data points and hence can be used to make further predictions about the value of the dependent variable where it is not known initially.

Least Square Method Formula

The least-square method states that the curve that best fits a given set of observations, is said to be a curve having a **minimum sum of the squared residuals (or deviations or errors)** from the given data points. Let us assume that the **given points of data are (x_1, y_1), (x_2, y_2), (x_3, y_3), …, (x_n, y_n) in which all x's are independent variables, while all y's are dependent ones.** Also, suppose that f(x) is the fitting curve and d represents error or deviation from each given point.

Now, we can write:

$d_1 = y_1 - f(x_1)$

$d_2 = y_2 - f(x_2)$

$d_3 = y_3 - f(x_3)$

.....

$d_n = y_n - f(x_n)$

The least-squares explain that the curve that best fits is represented by the property that the sum of squares of all the deviations from given values must be minimum, i.e:

$$S = \sum_{i=1}^{n} d_i^2$$

$$S = \sum_{i=1}^{n} [y_i - f_{xi}]^2$$

$$S = d_1^2 + d_2^2 + d_3^2 = ... + d_n^2$$

Least Square Method formula

Sum = Minimum Quantity

Suppose when we have to determine the equation of line of best fit for the given data, then we first use the following formula.

The equation of least square line is given by Y = a + bX

Normal equation for 'a':

$\sum Y = na + b\sum X$

Normal equation for 'b':

$\sum XY = a\sum X + b\sum X^2$

Solving these two normal equations we can get the required trend line equation.

Thus, we can get the line of best fit with formula y = ax + b

Solved Example

The Least Squares Model for a set of data $(x_1, y_1), (x_2, y_2), (x_3, y_3), \ldots, (x_n, y_n)$ passes through the point (x_a, y_a) where x_a is the average of the x_i's and y_a is the average of the y_i's. The below example explains how to find the equation of a straight line or a least square line using the least square method.

Question:

Consider the time series data given below:

x_i	8	3	2	10	11	3	6	5	6	8
y_i	4	12	1	12	9	4	9	6	1	14

Use the least square method to determine the equation of line of best fit for the data. Then plot the line.

Solution

Mean of x_i values = (8 + 3 + 2 + 10 + 11 + 3 + 6 + 5 + 6 + 8)/10 = 62/10 = 6.2

Mean of y_i values = (4 + 12 + 1 + 12 + 9 + 4 + 9 + 6 + 1 + 14)/10 = 72/10 = 7.2

Straight line equation is y = a + bx.

The normal equations are

$\sum y = an + b\sum x$

$\sum xy = a\sum x + b\sum x^2$

x	y	x^2	xy
8	4	64	32
3	12	9	36
2	1	4	2
10	12	100	120
11	9	121	99
3	4	9	12
6	9	36	54
5	6	25	30
6	1	36	6
8	14	64	112
$\Sigma x = 62$	$\Sigma y = 72$	$\Sigma x^2 = 468$	$\Sigma xy = 503$

Substituting these values in the normal equations,

$10a + 62b = 72$….(1)

$62a + 468b = 503$….(2)

$(1) \times 62 - (2) \times 10$,

$620a + 3844b - (620a + 4680b) = 4464 - 5030$

$-836b = -566$

$b = 566/836$

$b = 283/418$

$b = 0.677$

Substituting $b = 0.677$ in equation (1),

$10a + 62(0.677) = 72$

$10a + 41.974 = 72$

$10a = 72 - 41.974$

$10a = 30.026$

$a = 30.026/10$

$a = 3.0026$

Therefore, the equation becomes,

$y = a + bx$

$y = 3.0026 + 0.677x$

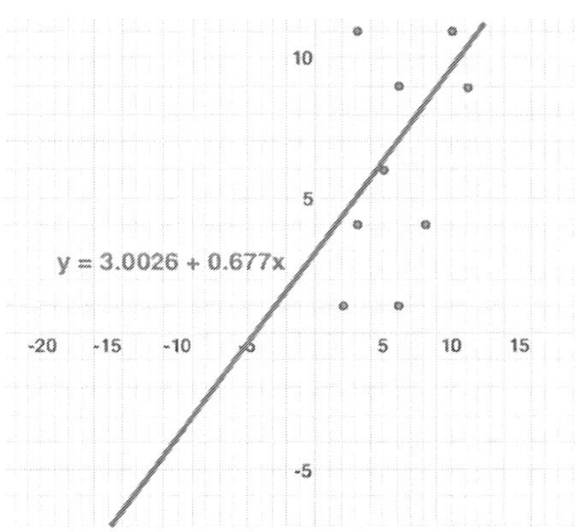

This is the required trend line equation. Now, we can find the sum of squares of deviations from the obtained values as:

$d_1 = [4 - (3.0026 + 0.677*8)] = (-4.4186)$

$d_2 = [12 - (3.0026 + 0.677*3)] = (6.9664)$

$d_3 = [1 - (3.0026 + 0.677*2)] = (-3.3566)$

$d_4 = [12 - (3.0026 + 0.677*10)] = (2.2274)$

$d_5 = [9 - (3.0026 + 0.677*11)] =(-1.4496)$

$d_6 = [4 - (3.0026 + 0.677*3)] = (-1.0336)$

$d_7 = [9 - (3.0026 + 0.677*6)] = (1.9354)$

$d_8 = [6 - (3.0026 + 0.677*5)] = (-0.3876)$

$d_9 = [1 - (3.0026 + 0.677*6)] = (-6.0646)$

$d_{10} = [14 - (3.0026 + 0.677*8)] = (5.5814)$

$\sum d^2 = (-4.4186)^2 + (6.9664)^2 + (-3.3566)^2 + (2.2274)^2 + (-1.4496)^2 + (-1.0336)^2 + (1.9354)^2 + (-0.3876)^2 + (-6.0646)^2 + (5.5814)^2 = 159.27990$

Limitations for Least-Square Method

The least-squares method is a very beneficial method of curve fitting. Despite many benefits, it has a few shortcomings too. One of the main limitations is discussed here.

In the process of regression analysis, which utilizes the least-square method for curve fitting, it is inevitably assumed that the errors in the independent variable are negligible or zero. **In such cases, when independent variable errors are non-negligible, the models are subjected to measurement errors**. Therefore, the least square method may even lead to hypothesis testing, where parameter estimates and confidence intervals are taken into consideration due to the presence of errors occurring in the independent variables.

Q1 How do you calculate least squares?

Let us assume that the given points of data are (x_1, y_1), (x_2, y_2), …, (x_n, y_n) in which all x's are independent variables, while all y's are dependent ones. Also, suppose that f(x) be the fitting curve and d represents error or deviation from each given point. The least-squares explain that the curve that best fits is represented by the property that the sum of squares of all the deviations from given values must be minimum.

Q2 How many methods are available for the Least Square?

There are two primary categories of least-squares method problems:

i) Ordinary or linear least squares

ii) Nonlinear least squares

Q3 What is the principle of least squares?

The least squares principle states that by getting the sum of the squares of the errors a minimum value, the most probable values of a system of unknown quantities can be obtained upon which observations have been made.

Q4 What does the least square mean?

The least square method is the process of obtaining the best-fitting curve or line of best fit for the given data set by reducing the sum of the squares of the offsets (residual part) of the points from the curve.

Q5 What is least square curve fitting?

The least-squares method is a generally used method of the fitting curve for a given data set. It is the most prevalent method used to determine the trend line for the given time series data.

Linear Models

A linear model is a model that assumes linearity in the system. It describes a relationship between two quantities that show a constant rate of change. A linear model also called linear regression model can be represented by an equation or a graph with a straight line. A linear model can be used to translate natural phenomena into comprehensible and recurring parameters.

1. Fixed Effect Models

In statistics, a fixed effects model is a statistical model in which the model parameters are fixed or non-random quantities. This is in contrast to random effects models and mixed models in which all or some of the model parameters are random variables. In many applications including econometrics and biostatistics, a fixed effects model refers to a regression model in which the group means are fixed (non-random) as opposed to a random effects model in which the group means are a random sample from a population. Generally, data can be grouped according to several observed factors. The group means could be modelled as fixed or random effects for each grouping. In a fixed effects model each group mean is a group-specific fixed quantity.

In panel data where longitudinal observations exist for the same subject, fixed effects represent the subject-specific means. In panel data analysis the term fixed effects estimator (also known as the within estimator) is used to refer to an estimator for the coefficients in the regression model including those fixed effects (one time-invariant intercept for each subject).

Qualitative description

Such models assist in controlling for omitted variable bias due to unobserved heterogeneity when this heterogeneity is constant over time. This heterogeneity can be removed from the data through differencing, for example by subtracting the group-level average over time, or by taking a first difference which will remove any time invariant components of the model.

There are two common assumptions made about the individual specific effect: the random effects assumption and the fixed effects assumption. The random effects assumption is that the individual-specific effects are uncorrelated with the independent variables. The fixed effect assumption is that the individual-specific effects are correlated with the independent variables. If the random effects assumption holds, the random effects estimator is more efficient than the fixed effects estimator. However, if this assumption does not hold, the random effects estimator is not consistent. The Durbin–Wu–Hausman test is often used to discriminate between the fixed and the random effects models.

Formal model and assumptions

Consider the linear unobserved effects model for N observations and T time periods:

$$Y_{it} = X_{it}\beta + \alpha_i + u_{it} \text{ for } t = 1, \ldots, T \text{ and } i = 1, \ldots, N$$

Where,

- Y_{it} is the dependent variable observed for individual i at time t
- $X_{it}\beta$ is the time-variant 1 x k (the number of independent variables) regressor vector.
- β is the k x 1 matrix of parameters.
- α_i is the unobserved time-invariant individual effect. For example, the innate ability for individuals or historical and institutional factors for countries.
- u_{it} is the error term.

Unlike X_{it}, α_i cannot be directly observed.

Unlike the random effects model where the unobserved α_i is independent of X_{it} for all t =1,………,T , the fixed effects (FE) model allows α_i to be correlated with the regressor matrix X_{it}. Strict exogeneity with respect to the idiosyncratic error term uit is still required.

Mixed model

A mixed model, mixed-effects model or mixed error-component model is a statistical model containing both fixed effects and random effects. These models are useful in a wide variety of disciplines in the physical, biological and animal sciences. They are particularly useful in settings where repeated measurements are made on the same statistical units (longitudinal study), or where measurements are made on clusters of related statistical units.

Fixed effects encapsulate the tendencies/trends that are consistent at the levels of primary interest. These effects are considered fixed because they are non-random and assumed to be constant for the population being studied. For example, when studying education, a fixed effect could represent overall school level effects that are consistent across all schools. Effect of climate or effect of a single sire use in a dairy farm is fixed effect on all progenies.

The Random Effect

A key component of the mixed model is the incorporation of random effects with the fixed effect. Fixed effects are often fitted to represent the underlying model. In Linear mixed models, the true regression of the population is linear, β. The fixed data is fitted at the highest level. Random effects introduce statistical variability at different levels of the data hierarchy. These account for the unmeasured sources of variance that affect certain groups in the data. For example, the differences between student 1 and student 2 in the same class, or the differences between class 1 and class 2 in the same school. Same sire but different cows in a firm create differences in progeny performances.

Ronald Fisher introduced random effects models to study the correlations of trait values between relatives. In the 1950s, **Charles Roy Henderson** provided best linear unbiased estimates (BLUE) of fixed effects and best linear unbiased predictions (BLUP) of random effects.

In matrix notation a linear mixed model can be represented as

$$y = X\beta + Zu + \varepsilon$$

Where,

- y is a known vector of observations, with mean $E(y)=X\beta$
- β is an unknown vector of fixed effects
- u is an unknown vector of random effects, with mean $E(u)= 0$ and variance–covariance matrix var(u)=G
- ε is an unknown vector of random errors, with mean $E(\varepsilon)=0$ and $var(\varepsilon)=R$
- X is the known design matrix for the fixed effects relating the observations y to β, respectively.
- Z is the known design matrix for the random effects relating the observations y to u, respectively.

Estimation

The joint density of y and u can be written as: f(y, u)= f(y/u) f(u). Assuming normality, $u \sim N(O, G)$, $\varepsilon \sim N(O, R)$ and Cov (u ε) = 0 and maximizing the joint density over β and u gives Henderson's mixed model equations (MME) for linear mixed models:

$$\begin{pmatrix} X'R^{-1}X & X'R^{-1}Z \\ Z'R^{-1}X & Z'R^{-1}Z+G^{-1} \end{pmatrix}\begin{pmatrix} \hat{\beta} \\ \hat{u} \end{pmatrix} = \begin{pmatrix} X'R^{-1}y \\ Z'R^{-1}y \end{pmatrix}$$

The solutions to the MME, $\hat{\beta}$ and $\hat{U}$ are best linear unbiased estimates and predictors for β and u, respectively. When the conditional variance is known, then the inverse variance weighted least squares estimate is best linear unbiased estimates. However, the conditional variance is rarely, if ever, known. So it is desirable to jointly estimate the variance and weighted parameter estimates when solving MMEs.

Mixed effects models combine both fixed effects and random effects. Let's investigate into what these terms mean and how they're used in the context of animal research:

Fixed Effects

In previous analyses using General Linear Models, we treated factors as fixed effects. But what exactly are fixed effects?

- Fixed effects represent systematic and predictable influences on the response variable. They are considered constant across all levels of the factor. e.g. effect of climate in a farm, effect of one sire in a farm is same to all progeny.

- For instance, if we're studying the effect of different diets on animal growth, the diet type would be a fixed effect. We assume that the diet has a consistent impact on growth across all individuals.
- However, fixed effects don't account for variability due to individual differences or other sources of randomness.

Random Effects

Random effects address sources of variance that are not systematic or predictable. They capture individual differences and other factors that introduce variability.

Here are two primary purposes for incorporating random effects:

- Modelling Non-Independence: When our study involves repeated measurements on the same subjects (e.g., measuring individuals within cages), we need to account for non-independence. Random effects (e.g., treating subject, family, or cage as random) help avoid pseudo-replication.
- Biological Research Questions: Biologists often use random effects to quantify variance due to genetic vs. environmental causes. For example:
- Quantifying genetic contributions to a phenotype.
- Studying phenotypic plasticity patterns within populations.
- Investigating individual differences in properties of specific cells.
- Unlike fixed effects, random effects allow us to address biological questions related to variation rather than just mean differences between groups.

Applications

- Quantitative Genetics: Random effects are commonly used to dissect genetic and environmental contributions to traits.
- Ecology: Researchers employ mixed effects models to account for non-independence in ecological data.
- Neuroscience: For instance, mixed effects models help and explore how among-animal differences contribute to variability in neuronal properties.
- Remember, mixed effects models are a vast subject, they provide a powerful tool for understanding complex biological systems and accounting for both systematic and random influences.

17

Understanding BLUEs, BLUPs and Breeding Values in Linear Mixed Models

- In statistics, best linear unbiased prediction (BLUP) is used in linear mixed models for the estimation of random effects. BLUP was derived by Charles Roy Henderson in 1950 but the term "best linear unbiased predictor" (or "prediction") seems not to have been used until 1962 (Henderson, 1975, 1985).
- **"Best linear unbiased predictions" (BLUPs) of random effects are similar to best linear unbiased estimates (BLUEs) of fixed effects.** The distinction arises because it is conventional to talk not about estimating fixed effects but rather about predicting random effects, but the two terms are otherwise equivalent.

BLUE and BLUP are both acronyms used in statistics

- BLUE stands for Best Linear Unbiased Estimates and refers to the solutions (or estimates) associated with the **fixed effects of a model.**
- BLUP stands for Best Linear Unbiased Predictions and refers to the solutions (but identified as predictions) **associated with the random effects of a model**.
- Both BLUEs and BLUPs are **linear functions of the data** and have the lowest variance.
- The expected value of a mean estimate for an individual **equals its true value** for BLUEs.
- BLUPs are similar to BLUEs, but it is conventional to talk about estimating fixed effects but predicting random effects.
- **BLUP is the standard selection method in animal breeding where the breeding values of sires are estimated based on progeny performance** to select superior genotypes and to breed superior families.

The fitting of a linear mixed model (LMM) can be divided into two steps. The first one is the estimation of the variance components. Once these variance components are estimated, we then go onto the second step, where these variances are used to obtain estimates of the **fixed effects and predictions of random effects.** This is where we need to differentiate between BLUEs and BLUPs. But what are these?

The distinction between BLUEs and BLUPs

Best Linear Unbiased Estimates (BLUEs) are the solutions (or estimates) associated with the fixed effects and Best Linear Unbiased Predictions (BLUPs) are the solutions (but identified as predictions) associated with the random effects of a model. But what does Best Linear Unbiased mean?

Best Among all possible unbiased linear estimators these solutions have minimum variance

Linear Solutions are formed from a linear combination of the observations

Unbiased Expectations of these solutions are equal to their true values

Henderson's MME and the animal model

The use of BLUPs to predict random effects was first described by C. R. Henderson. He developed the **mixed model equations** (MME, see equation [2]) for LMMs in order to calculate BLUPs of breeding values (or any random effect) and BLUEs of fixed effects.

Let's have a look at how solutions are obtained from a linear mixed model. For this, we will consider the model written in matrix notation:

[1] y=Xb+Zu+e

where y, b, u, and e are vectors of observations, fixed effects, random effects, and random residuals, respectively; and X and Z are design matrices connecting the observations to the effects.

The above model is the basis for most genetic analyses with what is known as the Animal Model, that once fitted provides us with an estimation of the breeding values for each of the 'animals' (individuals) considered in the model.

Considering the Animal Model on its matrix notation of above, we have **b corresponding to a series of nuisance fixed effects**, for example contemporary group, age or replicate. However, what is more interesting to us is **u, corresponding to the breeding values**. As this factor is random, we have distributional assumptions, namely that u ~ MVN $(0,\sigma^2 aA)$ where the **A**

is the numerator relationship matrix that describes **the additive genetic relationship between individuals**, and $\sigma^2 a$ is the additive variance. (MVN is multivariate normality test).

As the error or residual term, **e, is also a random effect, we have their distributional assumptions of e ~ MVN(0,σ^2eI)**, with I an identity matrix **and σ^2e the residual variance (or mean square error).**

As we can see, the central problem in predicting breeding values from observed phenotypic data is separating the genetic and environmental effects. This separation is clearly done by the Animal Model.

Solving the system of equations: Henderson's MME

So, in order to get our solutions (i.e., BLUEs and BLUPs), we need to solve the following system of equations, known as Henderson's MME:

[2]

$$\begin{vmatrix} X'X & X'Z \\ Z'X & Z'Z+A^{-1}a \end{vmatrix} \begin{vmatrix} b \\ \mu \end{vmatrix} = \begin{vmatrix} X'y \\ Z'y \end{vmatrix}$$

All the elements we have previously defined, except for $a=\sigma^2 e/\sigma^2 a$, which is formed from variance components estimated in the first step when fitting a LMM. This system can involve hundreds or even thousands of equations, requiring some intense computational calculations in order to estimate our solutions.

A different way to see the above equations is to express **the formulae for BLUEs as**

[3]

$$b = (X'V^{-1}X)^{-}X'V^{-1}y$$

and **for the BLUPs (breeding values) as**

[4]

$$u = \sigma^2 aAZ'V^{-1}(y-Xb)$$

where $V=\sigma^2 aZAZ'+\sigma^2 eI$

Equivalence to Weighted Least Squares (WLS)

Interestingly, expression [3] is equivalent to the estimation of Weighted Least Squares (WLS) as described for linear regression or linear models. Here, the weights are defined as

$$W = V^{-1}$$

This is relevant as the elements on V (particularly the diagonals) will be associated with the amount of information, or weight, that each record has (in this case phenotypic response of an individual).

Breeding values and the breeder's equation

Also, the expression for the breeding values is extremely relevant. Note that [4] can be written as:

$u = H(y - Xb)$

where H is $\sigma^2 aAZ'V^{-1}$ **resembling the calculation of heritability** with the genetic component in the numerator ($\sigma^2 aAZ'$) and the phenotypic variance (V) in the denominator as an inverse. Also, y−Xb, **represents the phenotypic response 'corrected' for nuisance effects**. Therefore, the above **expression resembles the breeder's equation:**

$gi = h^2\, Si = h^2\, (yi - \mu)$

where the **index i is used to identify the ith *h* individual.**

Therefore, fitting a LMM using an Animal Model is equivalent to that well-known expression, but in this case, the specification of matrices connecting pieces of data allows us to use all information and relationships between individuals and to have different weights for each of these pieces. **Therefore, Henderson's MME provide us with the best linear unbiased predictions of our breeding values. (Nazarian and Gezan, 2016).**

In Summary

- **BLUEs help estimate fixed effects**, while **BLUPs predict random effects** (such as breeding values) in animal breeding models.
- Henderson's work laid the foundation for understanding and applying these concepts, allowing us to make informed decisions in animal breeding.

BLUPF90 programs

Solving of mixed model equations to get BLUPs and BLUEs was the original purpose of the BLUPF90 programs. These are stored in memory and solutions can be obtained by direct inversion or by iterative methods. The default method is the preconditioned conjugate gradient algorithm. Pedigree additive relationships (with or without unknown parent groups) are supported, and these can be combined with genomic information by adding extra relationship matrices following the theory of **single-step genomic best linear unbiased**

prediction (ssGBLUP). Other (co)variance structures can be defined including: parental dominance, metafounders, relationships derived for honey bee production and also user defined matrices, e.g. for **genomic epistasis or additive-dominance with inbreeding or an user defined matrix**. Prediction error (co)variances can be obtained from the sparse inverse of the MME, to derive accuracy for animal models or for random regression models. BLUPF90 supports heterogeneous residual variances as used in GIBBS3F90 where the user can supply residual variance components as a file *(PDF)* **BLUPF90 suite of programs for animal breeding with focus on genomics.** Available from: https://www.researchgate.net/publication/338122732_BLUPF90_suite_of_programs_for_animal_breeding_with_focus_on_genomics (Lourenco *et al.*, 2020).

Distinction between linear models and mathematical models

Mathematical Model

A mathematical model is an abstract representation of a real-world system using mathematical concepts and language.

It serves as a tool to describe, analyse, and predict the behaviour of various systems across different domains, including applied mathematics, natural sciences (such as physics, biology, and chemistry), engineering, and even social sciences (like economics and psychology). Mathematical models can take various forms, such as:

Dynamical systems: Describing how quantities change over time.

Statistical models: Analyzing data and making predictions.

Differential equations: Representing relationships involving rates of change.

Game theoretic models: Studying strategic interactions.

These models help us understand complex phenomena, explore the effects of different components, and make informed predictions.

Linear vs. Nonlinear Models

Linear Model

A mathematical model is considered linear if all its operators exhibit linearity.

Linearity implies that the model's relationships can be expressed as linear combinations of variables.

Examples

- In a statistical linear model, the relationship between parameters is assumed to be linear, even if predictor variables themselves are nonlinear.
- A differential equation is linear if it involves linear differential operators, even if it contains nonlinear expressions.
- In mathematical programming, a model is linear if both objective functions and constraints are represented entirely by linear equations.

Nonlinear Model

If a model doesn't meet the criteria for linearity, it's considered nonlinear. Nonlinear models allow for more complex relationships and interactions.

Examples:

- Systems with exponential growth or decay.
- Nonlinear optimization problems.
- Chaotic dynamical systems.

The distinction between linear and nonlinear models is context-dependent, and sometimes linear models may contain nonlinear elements.

In summary, mathematical models provide powerful tools for understanding and predicting real-world phenomena, while linear models specifically adhere to linear relationships between variables.

Aspect	**Linear Models**	**Mathematical Models**
Definition	Linear models exhibit linearity in their structure. The relationship between variables is linear.	Mathematical models encompass a broader category. They can be linear or nonlinear, depending on the form of the equations.
Equation Example	Example: (y = ax1 + bx2), where both the model and parameters (a) and (b) are linear.	Example: (y = a sin(x1) + bx2), which is nonlinear in the model but linear in its parameters.
Characteristics	Simplicity and straightforwardness. Assumes a direct proportionality between variables.	Greater flexibility. Allows for more complex relationships between variables.

Aspect	Linear Models	Mathematical Models
Parameter Linearity	Linear models have linear parameters (e.g., (a) and (b)).	Nonlinear models may have linear or nonlinear parameters.
Common Applications	Widely used for basic statistical analysis. Simple regression models fall into this category.	Applied in various fields, including physics, engineering, economics, and more.

18

Animal Genetics and Breeding Question Bank for B.V.Sc. & A.H.

Subject: Biostatistics and computer application (UNIT-I)

Write true (T) or false (F)

Q1. Write true (T) or false (F)

1. The mean and mode are always equal in case of normal distribution.
2. Pie diagram is circular diagram.
3. Skewness is the measure of asymmetry of distribution.
4. Mean sum of squares are the variances.
5. Coefficient of variation may be utilized for comparing two data of different unit.
6. Goodness fit is tested by t-test.
7. Variance is a measure of central tendency.
8. CRD can be use where experimental material is heterogenous.
9. Frequency polygon is made by joining the middle points of the top of each rectangle of histogram.
10. Correlation coefficient has no unit.
11. Any two events A and B are said to be independent if P(AB)=P(A).P(B).
12. Range can be calculated when mean and median are known.
13. F-test is generally used to test significance of two variances.
14. The range of correlation coefficient is -1 to +1.

15. In Poisson distribution the mean and variance are same.
16. RBD gives large error if blocks are not homogenous.
17. Mode is not usually affected by extreme values.
18. Mode is the best measure of central tendency.
19. Level of significance (α) is the probability of committing type I error.
20. Differences among r and byx, bxy in two populations are tested by t-test.
21. RBD is more accurate than CRD in most type of experimental work.
22. RBD is inappropriate when experimental units are heterogeneous.
23. Sum of deviations from the mean is equal to zero.
24. Paired t-test is applied when the two samples are dependent.
25. Mode is the best measure of central tendency.
26. Rejecting a null hypothesis (H0) when it is true is called type II error.
27. Level of significance (α) is the probability of committing type I error.
28. Pie diagram is used to depict the proportions of different classes.
29. The product of regression coefficient of y on x (byx) and regression coefficient of x on y (bxy) is always positive.
30. Mode is not usually affected by extreme values.
31. SPSS computer software is used for data analysis.
32. F-test is generally used for testing the significance of two variances.
33. The range of correlation coefficient is 0 to +1.
34. Mode is the best measure of central tendency.
35. Rejecting a null hypothesis (H0) when it is true is called type-ll error.
36. Regression is dimensionless quantity.
37. Milk yield is an example of discontinuous trait.
38. Sum of deviations from mean is equal to zero.
39. Regression coefficient of Y on X and X on Y are always equal.
40. Standard deviation is a measure of dispersion.
41. Variance is independent of origin.

42. Regression coefficient lies between -1 and +1.
43. ED50 and LD50 are same.
44. Histogram and frequency polygon are same.
45. Type I error is more serious than Type II error.
46. In binomial distribution mean and variance are equal.
47. Positional measure of central tendency is median.
48. Probability value ranges from 0 and 1.
49. Co-efficient of variation is a relative measure of dispersion.
50. RBD more precise than CRD.

Multiple Choice Questions

Q.2. Multiple Choice Questions

1. Test of significance of regression coefficient can be test with,

   a. t-test　　b. F test

   c. Both t and F test　　d. Chi-square test.

2. Presence of linkages in dihybrid inheritance can be tested by,

   a. Z- test　　b. F- test

   c. Chi-square test　　d. t-test.

3. Which of the following is known as standard normal deviate test of significance?

   a. t-test　　b. F-test

   c. Chi-Square　　d. Z-test.

4. The value of chance (or probability) that is used to provide lines of demarcation between acceptance and rejection of null hypothesis is known as

   a. Level of significance　　b. Confidence interval

   c. Critical region　　d. Hypothesis.

5. Mean and variance in poisson distribution is estimated to be

   a. np and nq respectively　　b. np and np respectively,

   c. npq and nq only　　d. nq and np respectively.

6. The father of statistics is

   a. R. A. Fisher
   b. Gauss
   c. Karl Pearson
   d. None of these.

7. The unknown variation caused by extraneous factors which are beyond the control of experimenter is called,

   a. Treatment
   b. Experimental error
   c. Experimental unit
   d. Yield.

8. The measure of variation that is least affected by extreme observation is

   a. Mean deviation
   b. Quartile deviation
   c. Standard deviation
   d. Range

9. Which of the following % of items in a normal distribution is covered by mean $\pm 2\sigma$

   a. 95.54
   b. 95.45
   c. 96.45
   d. 99.73

10. The test statistic used to test equality of mean of two independent small samples with known population variance is

    a. F-test
    b. Z-test
    c. t-test
    d. X^2 -test

11. In which of the following statistical test expected frequency of each class should be greater than 5

    a. F-test
    b. Z-test
    c. t-test
    d. X^2 -test

12. Random allotment of treatments to different experimental unit is called

    a. Replication
    b. Local control
    c. Randomization
    d. None of the above.

13. The term "regression" was coined by ---------------------------------in the 19th century to describe a biological phenomenon.

    a. Francis Galton
    b. R.A. Fisher
    c. Karl Pearson
    d. None of above

14. The student distribution or t test was given by

a. Karl Pearson b. Snedecor
c. W.S. Gosset d. RA Fisher

15. Chi square test was developed by

a. Karl Pearson b. Snedecor
c. W.S. Gosset d. RA Fisher

16. Variance is expressed as deviation from

a. Mode b. Median
c. Mean d. Variable

17. Coefficient of correlation was given by

a. Karl Pearson b. Snedecor
c. W.S. Gosset d. RA Fisher

18. Relative measure of dispersion is

a. Mean deviation b. Standard deviation
c. Both a and b d. Coefficient of variation

19. Measure of dispersion include

a. Mean deviation b. Standard deviation
c. Range d. All of the above

20. Which of the following type of error is used for rejecting null hypothesis (Ho) when it is true:

a. Type I error b. Type II error
c. Both a& b d. None of the above

21. Standard normal distribution has mean equal to

a. One b. Two
c. Zero d. None of the above.

20. In RBD the data is classified

a. One way b. Two way
c. Three way d. More than that

21. The best estimate of central tendency in normal distribution

a. Geometric mean
b. Arithmetic mean
c. Harmonic mean
d. None of above

22. A line graph of a cumulative frequency or cumulative relative frequency distribution is known as

a. Ogive
b. Cumulative Polygon
c. Both 1 and 2
d. Frequency curve

23. The value of the χ^2 will always be

a. Negative
b. Positive
c. Depends on number of observations.
d. Can take any value.

24. Critical difference is used for the identification of

a. Comparison of treatment means
b. Used in chi-square test.
c. Used in t-test
d. None of the above.

25. The regression equation of Y on X can be written as

a. $\hat{y} = y + bx$
b. $\hat{y} = y(bx)$
c. $\hat{y} = ybx$
d. $\hat{y} = y/bx$

28. Primary data are:

a. raw original unprocessed data
b. based on choice of enumerator
c. statistical manipulation.
d. Obtained from data base.

29. Critical difference test is used to

a. compare subclass means
b. compare two variances.
c. compare one mean of one effect.
d. It is done after chi-square test.

30. Which of the following biological process follow the law of probability?

a. Gamete formation
b. Hormone secretions
c. Metabolic process
d. Metamorphosis.

31. The two line of regression same at

a. $r = 0.5$
b. $r = 0.0$
c. $r = 0.89$
d. $r = \pm 1$

32. The repetition of treatments under investigation is known as
    a. Randomization
    b. Local control
    c. Replication
    d. None of the above.
33. Parameters of binomial distributions are:
    a. n and p
    b. p and q
    c. np and npq
    d. None of the above
34. Accepting a true hypothesis is:
    a. Type-l error
    b. Type-ll error
    c. any a or b
    d. None of the above
35. RBD is the simplest design where:
    a. All basic principles of design are used
    b. Only replication is used
    c. Replication and randomization is used
    d. Randomization is used
36. Central value of z for two tail test at 0.05 level of significance is:
    a. -1.96 and 1.96
    b. -2.81 and 2.81
    c. -2.58 and 2.58
    d. None of the above
37. When "n" is an odd number then median is defined as;
    a. Most repeated value
    b. Median of two middle values
    c. Sum of the values
    d. Middle value
38. For a group data the class interval having maximum frequency is known as
    a. Median class
    b. Modal class
    c. Median
    d. Mode
39. A sample is called small when N is:
    a. Greater than 30
    b. Less than 30
    c. Both of the above
    d. None of the above

40. Critical region of Z for two tail test at 0.05 level of significance is:

a. -1.96 and +1.96
b. -2.81 and +2.81
c. -2.58 and +2.58
d. None of the above

41. The standard deviation of 4, 4, 4 and 4 is

a. 0
b. 4
c. 8
d. 16

42. Test of significance includes:

a. t-test
b. chi-square test
c. F-test
d. All of the above

43. Which average is much affected by extreme observations:

a. Mode
b. Median
c. Arithmetic mean
d. None of the above

44. Chi square test is a:

a. measure of central tendency
b. measure of dispersion
c. measure of correlation
d. test of goodness of fit

45. In poisson distribution the mean is equal to:

a. variance
b. twice the variance
c. half of variance
d. standard deviation

46. If 15 is subtracted from each observation the standard deviation of data will be:

a. increased by 15
b. decreased by 15
c. increased by 25
d. same

47. Covariance a measure of the joint variance of two variables may be:

a. positive
b. negative
c. either positive or negative
d. None of the above

48. One of the methods of determining mode is:

a. 2 median -3 mean
b. 3 median + 2 mean
c. 3 median- 2 mean
d. 2 median + 3 mean

49. Variance is expressed as deviation from:

a. Mode
b. Median
c. Mean
d. Variable

50. If there are 6 treatments having 4 replication each then degree of freedom for treatment is

a. 2
b. 5
c. 6
d. 3

Q. 3. Fill in the blanks

1. ---------------------- is the science of collection, presentation, analysis and interpretation of numerical data from the logical analysis.
2. A ------------- is a circle divided into portions that represent the relative frequencies or percentages of a population or a sample belonging to different categories.
3. The --------------- is also called as the second moment of dispersion.
4. The value of chance (or probability) that is use to provide lines of demarcation between acceptance and rejection of null hypothesis is known as ------------------------.
5. ---------------test is applied for comparison, when sample size is small and population parameter is known.
6. When two samples are related with each other, the mean of these samples can be compared using -------------------------- test statistics.
7. The -------------------------- coefficient measures the strength of the linear relationship between two variables.
8. The ---------------------- coefficient can be defined as the measure of change in the dependent variable per unit change in the independent variable.
9. To know the significant difference between treatment means in ANOVA which statistical test is applied ------------------------------------.
10. Bar diagrams are ------------------------- dimensional diagrams.
11. The ----------------- coefficient can be defined as the measure of change in the dependent variable per unit change in the independent variable.
12. To know the significant difference between treatment means in ANOVA which statistical test is applied ------------------------------.

13. Bar diagrams are ------------------- dimensional diagrams.
14. Chi square test was developed by ---------------------.
15. Median is a type of --------------------------- average.
16. In a positively skewed distribution, the value of mean is------------------- than the mode.
17. The measure of variation that is always expressed in percentage is ------ -----------.
18. Grouping of heterogenous experimental materials into block to make them homogenous is called ------------------------------.
19. The degree of association between the two variables is called ------------ ------.
20. ....is called Father of Biostatistics.
21. ....is called father of Statistics.
22. The skewness of a symmetric distribution is ....
23. The ....is the degree of asymmetry of distribution.
24. The relationship of correlation coefficient and regression coefficient is ....
25. In ANOVA experimental errors are assumed to be normally and independently distributed with mean .... and .... variance.
26. Probability of getting a male calf in a single calving is ....
27. As number of observations increases the t distribution tends to approaches ....
28. The limit of probability is........................
29. Kurtosis indicates .... of the curve.
30. The full expression of LD50 is ....
31. In Poisson distribution mean and variance are ....
32. ANOVA was developed by ....
33. The full expression of LD50 is ....
34. The unit of mean and SD are ....

35. In practice level of significance of ....is customary.
36. ....distribution is continuous probability distribution.
37. Milk production and fat percentage is an example of .... correlation.
38. The simplest possible measure of dispersion is ....
39. Square root of variance is ....
40. Chance of success or failure is ....
41. Normal distribution shows ....
42. Where there is perfect positive correlation, the value of correlation will be......................
43. Variance is .... of standard deviation.
44. Replication along with local control reduces the ....
45. In normal distribution mean, median and mode ....
46. The process of selecting individual from a population known as ....
47. Probability of two female calves in two successive calving is ....
48. The population parameters are denoted by using .... alphabets.
49. Body coat colours in cattle, wool type in sheep and comb pattern in poultry are example of ....trait/character.
50. Discrete data follow ....
51. Fourth moments about arithmetic mean measure ....

Q4. Write notes on following

1. Define Descriptive biostatistics.
2. Define Sample and Population.
3. Define frequency distribution and class interval.
4. Define frequency curve and frequency polygon.
5. Draw a normal distribution curve with different values of σ.

6. Describe correlation coefficient and regression coefficient.
7. What are the different sampling methods? Give a brief summary of 20th livestock census.
8. What are the different between primary and secondary data.
9. Define arthritic mean and write down merit and demerits of arthritic mean.
10. Describe in details the types of diagrams
11. Enumerate merits and demerits of CRD and RBD.
12. What do you mean by dispersion? What are various measures of dispersion? Describe about absolute mean deviation and coefficient of variation.
13. Describe standard deviation with its merits and demerits. 6 marks
14. Differentiate between paired and unpaired 't' test with appropriate formula.
15. What is Standard deviation? Give its computation formula.
16. What is the relationship between mean, median and mode?
17. Define standard error.
18. Define type l and type ll error.
19. Describe in brief.
    a) Describe Methods for collection of data.
    b) Enlist the importance of frequency distribution in data handling.
    c) What is the test hypothesis? Define null and alternative hypothesis.
    d) What are the measures of shape of distribution? Explain briefly about skewness of data.
    e) Enumerate merits and demerits of CRD and RBD.
    f) Define moments?
    g) What is t-test? Who discovered it?
20. Test of a Population Mean (Z test) versus t-test:
21. Relationship between arithmetic, geometric and harmonic mean.
22. Bayes' theorem

23. Critical Difference (C.D.).
24. What is binomial distribution?
25. What is poisson distribution?
26. What is chi-square distribution?
27. Define rank correlation?
28. What are the different types of bioassays? Define LD50, ED50, LC50 and Therapeutic index.
29. What is confidence interval?

Q5. Match The Following

1. Match column A and Column B

Sl. No.	Column A	Answer		Column B
1	Bar diagram	1. h	a	a. Variance
2	Attribute	2. i	b	b. Skewness
3	Regression	3. g	c	C. Cumulative frequency in percentage
4	When null hypothesis is true but it rejected	4. j	d	d. R. A. Fisher
5	Second moment	5. a	e	e. -1 to +1
6	ANOVA	6. d	f	f. Circular diagram
7	Lorenz curve	7. c	g	g. Francis Galton
8	Pie diagram	8. f	h	h. One dimensional
9	Third Moment	1. b	i	i. Qualitative character
10	Correlation	2. e	j	j. Type 1 error

2. Match column A and Column B

Sl. No.	Column A	Answer		Column B
1	Father of Biostatistics	1. h	a	James Bernoulli
2	Father of Statistics	2. i	b	1900
3	Correlation coefficient	3. j	c	1908
4	Normal distribution	4. g	d	W.S. Gosset
5	Student's t-test	5. d	e	1918
6	Binomial distribution	6. a	f	Simeon Denis Poisson in 1837
7	Chi square distribution given on	7. b	g	Abraham de Moivre in 1733
8	Student's t-test given on	8. c	h	Francis Galton
9	ANOVA was developed by Ronald Fisher in	9. e	i	Ronald Aylmer Fisher
10	Who discovered Poisson distribution?	10. f	j	Karl Pearson

AGB Answer Book

Q1. Write true (T) or false (F)

1. True **2.** True **3.** True **4.** True **5.** True **6.** False **7.** False **8.** False **9.** True **10.** True **11.** True **12.** False **13.** True **14.** True **15.** True **16.** Ture **17.** True **18.** False **19.** True **20.** True **21.** True **22.** True **23.** True **24.** True **25.** False **26.** False **27.** True **28.** True **29.** True **30.** True **31.** True **32.** True **33.** False **34.** False **35.** False **36.** False. **37.** False **38.** True **39.** False **40.** True **41.** True **42.** False 43. False **44.** False **45.** False **46.** False. **47.** True **48.** True **49.** True **50.** True.

Multiple Choice Questions

Q.2. Multiple Choice Questions

1. c. **2.** c. **3.** d. **4.** a. **5.** b. **6.**a. **7.**b. **8.** b. **9.** b. **10.** c. **11.** d. **12.** c. **13.** a. **14.** c. **15.** a. **16.** c, **17.** a. **18.** d. **19.** d. **20.** a. **21.** c. **22.** b. **23.** b. **24.** c. **25.** b. **26.** a. **27.** a. **28.** a **29.** a **30.** a. **31.** d **32.** c **33.** c. **34.** d. **35.** a. **36.** a. **37.** d. **38.** b. **39.** b. **40.** a. **41.** a. **42.** d. **43.** c. **44.** d. **45.** a. **46.** d. **47.** c. **48.** c. **49.** c. **50.** b.

Fill in the blanks

Q. 3. Fill in the blanks

1. Statistics/ Biostatistics 2. Pie diagram 3. Variance 4. Level of significance 5. t-test 6. Paired t-test 7. Correlation 8. Regression 9. F-test 10. one. 11. Regression 12. F-test 13. One 14. Karl Pearson 15. Positional 16. Greater 17. Coefficient of Variation (CV) 18. Randomized Blocks Design (RBD) 19. Correlation 20. Francis Galton 21. R. A. Fisher 22. Zero 23. Skewness 24. r^2 = byx.bxy 25. Zero and Constant 26. ½ or 50%. 27. Z distribution. 28. 0-1 29. Peakedness 30. lethal dose 50 31. equal, (both is λ =np).

32. Ronal Fisher 33. Lethal dose 50 34. Same 35. 0.05 or 0.01 36. Normal 37. Negative 38. Range 39. Standard deviation 40. Probability 41. Continuous distribution 42. +1 43. Square 44. Experimental error. 45. Coincide 46. sampling 47. ¼. 50. Greek 49. Qualitative 50. Binomial/Poisson/Chi square distribution 51. Kurtosis

Q4. Write short notes.

1. Define Descriptive biostatistics

Descriptive Biostatistics. We use these techniques to reduce a data set to manageable proportions, summarizing the trends and tendencies within it, in order to represent the results clearly. From these procedures we can produce **diagrams, tables and numerical descriptors**. Numerical descriptors include measures that convey where the **center of the data set lies, like the arithmetic mean or median, and measures of the scatter or dispersion of the data, such as the variance or range**.

2. Define Sample and Population.

Populations and samples

The population may be either a **real (or finite) group** or a **hypothetical (or infinite) group**. For example, if we are interested in the milk yield of cattle in Sahiwal, then the population is all cattle in Sahiwal. This is a real or finite population.

A population includes all representatives of a particular group, whereas a **sample** is a subgroup drawn from the population. We aim to choose a sufficiently large sample in such a manner that it is representative (i.e. is typical) of the population.

3. Define frequency distribution and class interval.

A frequency distribution shows the frequencies of occurrence of the observations in a data set. Often the distribution of the observed data is called an empirical frequency distribution.

The **class interval** is a term used in statistics when we are given a continuous series. Class means a group of numbers in which items are placed such as 0-10, 10-20, 20-30, etc. Class interval refers to the numerical width of any class in a particular distribution.

4. Define frequency curve and frequency polygon.

Polygon a graph formed by joining the midpoints of the tops of successive bars in a histogram with straight lines is called a *polygon*.

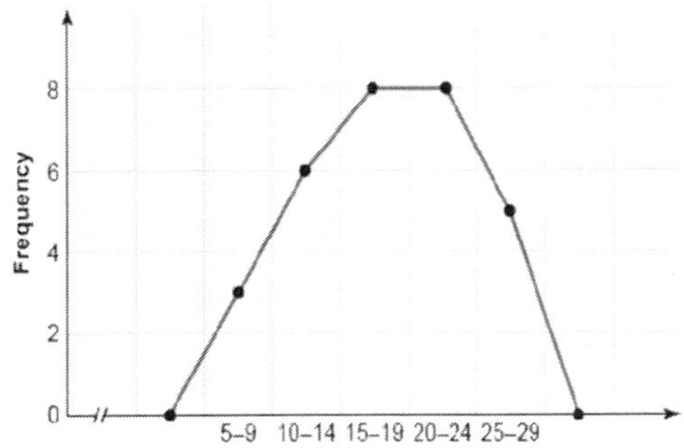

For a very large data set, as the number of classes is increased (and the width of classes is decreased), the frequency polygon eventually becomes a smooth curve. Such a curve is called a *frequency distribution curve* or simply a *frequency curve*.

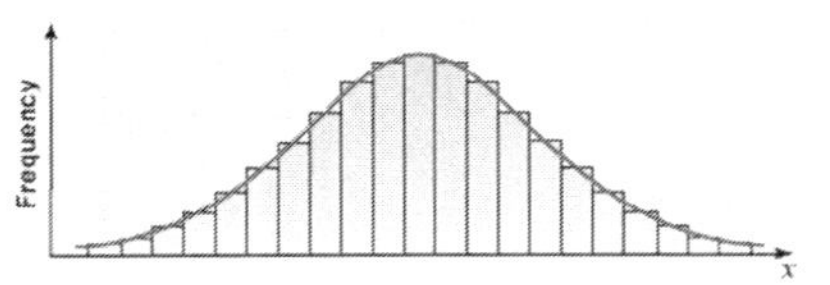

5. Draw a normal distribution curve with different values of σ.

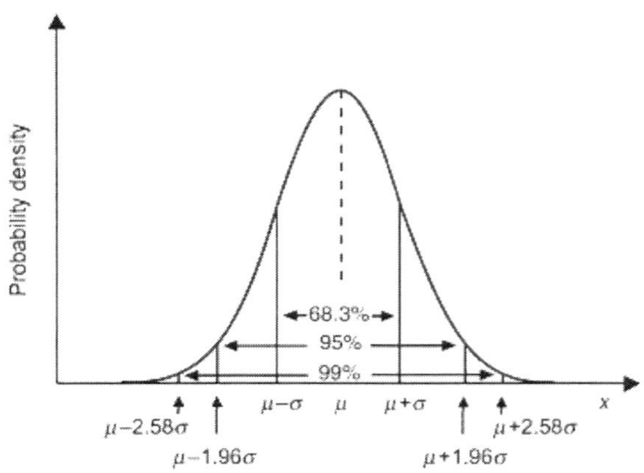

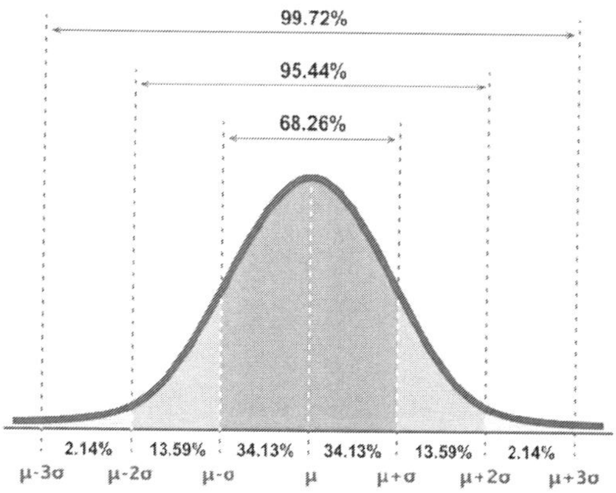

Areas under the Normal curve; the random variable, x, has mean = μ and standard deviation = σ

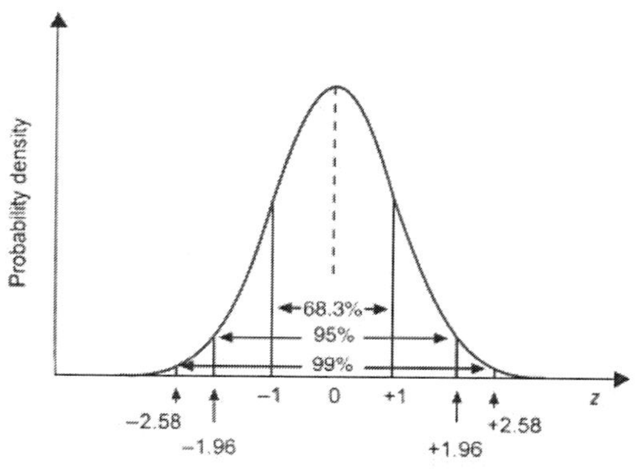

The Standard Normal curve; the random variable, z = (x – μ)/σ, has mean = 0 and standard deviation = 1

6. Describe correlation coefficient and regression coefficient.

Correlation coefficient:

We measure the degree of association between two variables by calculating **Pearson's product moment correlation coefficient**, usually just called the **correlation coefficient** or, sometimes, the **linear correlation coefficient**. It can take any value from −1 to +1.

If we take a random sample of n independent pairs of observations {($x1$, $y1$), ($x2$, $y2$), ($x3$, $y3$), . . . (xn, yn)} on two numerical variables, x and y, then we estimate the correlation coefficient, ρ (the Greek letter rho), in the population by the sample correlation coefficient

$$\rho X,Y = \frac{\text{cov}(X,Y)}{\sigma X \sigma Y}$$

Regression: can be defined as the measure of change in the dependent variable y per unit change in the independent variable x.

The regression coefficient described as followed,

$$b_{yx} = \frac{\text{Cov.xy}}{\text{Var.x}} = \frac{\Sigma XY - \Sigma X \Sigma Y / N}{\Sigma X2 - (\Sigma X)2 / N}$$

Likewise,

$$b_{xy} = \frac{\text{Cov.xy}}{\text{Var.}y} = \frac{\Sigma XY - \Sigma X \Sigma Y / N}{\Sigma Y2 - (\Sigma Y)2 / N}$$

and,

$$b_{yx.} b_{xy} = \frac{\text{Cov.xy}}{\text{Var.x}} = \frac{\text{Cov.xy}}{\text{Var.}y} = r^2$$

7. What are the different sampling methods? Give a brief summary of 20th livestock census.

Different method of sampling

1. **Random or Probability Sampling:** In this method, all the eligible individuals have a chance of selecting the sample from the whole sample space.

   a) **Simple Random Sampling:** As the sample size is large, and the item is chosen randomly, it is known as "Representative Sampling".

   b) **Restricted random sampling:**

      i) **Systematic sampling**: In the systematic sampling method, the items are selected from the target population by selecting the random selection point and selecting the other methods after a **fixed sample interval.**

      ii) **Stratified sampling**: In a **stratified sampling** method, the total population is divided into smaller groups to complete the sampling process. The small group is formed based on a few characteristics in the population. **After separating the population into a smaller group, the statisticians randomly (not all strata is selected) select the sample**.

iii) **Cluster sampling**: This method uses simple random sampling for the cluster of population. We break the population into many groups, then randomly **choose whole groups**.

c) **Non-random or Non-probability Sampling:** The non probability sampling method is a technique in which the researcher selects the sample based on subjective judgment rather than the random selection.

i) **Convenience Sampling:** In a convenience sampling method, the samples are selected from the population directly because they are **conveniently available for the researcher**.

ii) **Purposive or Judgmental Sampling:** In purposive sampling, the samples are selected only **based on the researcher's knowledge**. As their knowledge is instrumental in creating the samples, there are the chances of obtaining highly accurate answers with a minimum marginal error.

iii) **Quota Sampling:** In the quota sampling method, the researcher forms a sample that involves the individuals to represent the population based on **specific traits or qualities**.

iv) **Snowball sampling:** In this method, the samples have traits that are difficult to find. So, each identified member of a population is asked to find the other sampling units.

Give a brief summary of 20th livestock census

It was designed to capture Breed wise number of animals and poultry birds. Dog, Rabbit and elephant were included for first time.

i. The total Livestock population is 535.78 million

ii. The total number of cattle in the country is 192.49 million(^) (West Bengal)

iii. The total buffaloes in the country is 109.85 million (^) (Uttar Pradesh)

iv. The total sheep in the country is 74.26 million (^) (Telangana)

v. The Goat population in the country in 2019 is 148.88 million (^)

vi. The total Pigs in the country is 9.06 million(v)•

vii. The total Horses and Ponies in the country is 3.4 Lakhs (v)

viii. The total Poultry in the country is 851.81 million (^)

8. What are the different between primary and secondary data?

Data obtained from the population that originally collected by investigator are called **primary data**. If we obtain data from a source, or one collect primary data from a source is called as the primary source. Data obtained from an available source are called **secondary data**. The source from which we collect secondary data is known as secondary source.

In statistical analysis, collection of data plays a significant part. The method of collecting information is divided into two different sections, namely primary data and secondary data.

Primary data definition

Primary data is the data that is collected for the first time through personal experiences or evidence, particularly for research. It is also described as raw data or first-hand information. The mode of assembling the information is costly, as the analysis is done by an agency or an external organization, and needs human resources and investment. The investigator supervises and controls the data collection process directly. The data is mostly collected through observations, physical testing, mailed questionnaires, surveys, personal interviews, telephonic interviews, case studies, and focus groups, etc.

Secondary data definition

Secondary data is a second-hand data that is already collected and recorded by some researchers for their purpose, and not for the current research problem. It is accessible in the form of data collected from different sources such as government publications, censuses, internal records of the organization, books, journal articles, websites and reports, etc. This method of gathering data is affordable, readily available, and saves cost and time. However, the one disadvantage is that the information assembled is for some other purpose and may not meet the present research purpose or may not be accurate.

The differences between the primary and secondary data

	Primary Data	**Secondary Data**
Definition	Primary data are those that are collected for the first time.	Secondary data refer to those data that have already been collected by some other person.
Originality	These are original because these are collected by the investigator for the first time.	These are not original because someone else has collected these for his own purpose.
Nature of Data	These are in the form of raw materials.	These are in the finished form.
Reliability and Suitability	These are more reliable and suitable for the enquiry because these are collected for a particular purpose.	These are less reliable and less suitable as someone else has collected the data which may not perfectly match our purpose.
Time and Money	Collecting primary data is quite expensive both in the terms of time and money.	Secondary data requires less time and money; hence it is economical.
Precaution and Editing	No particular precaution or editing is required while using the primary data as these were collected with a definite purpose.	Both precaution and editing are essential as secondary data were collected by someone else for his own purpose.

9. Define arthritic mean and write down merit and demerits of arthritic mean.

(a) Arithmetic mean

The most commonly used measure of central tendency is the **arithmetic mean** (usually abbreviated to the **mean**). It is obtained by **adding together the observations in a data set and dividing by the number of observations in the set**. If the continuous variable of interest is denoted by *x* and there are *n* observations in the sample, then the sample mean (pronounced *x* bar) is

Arithmetic Mean

$$Mean(X) = \sum_{i=1}^{n} \frac{x_i}{n} = \frac{x_1 + x_2 + \ldots + x_n}{n}$$

$$ArithmeticMean = \frac{3+2+6+5}{4} = 4$$

Merits of Arithmetic Mean

I. The arithmetic mean is simple to understand and easy to calculate.

II. It is influenced by the value of every item in the series.

III. A.M is rigidly defined.

IV. It has the capability of further algebraic treatment.

V. It is a measured value and not based on the position in the series.

Demerits of Arithmetic Mean

I. It is changed by extreme items such as very small and very large items.

II. It can rarely be identified by inspection.

III. In some cases, A.M. does **not represent the original item**. For example, the average number of patients admitted to a hospital are **10.7 per day**.

IV. The arithmetic mean is not suitable in **extremely asymmetrical distributions**.

10. Describe in details the types of diagrams

1. One dimensional diagram

   Line diagram

   Bar Diagram

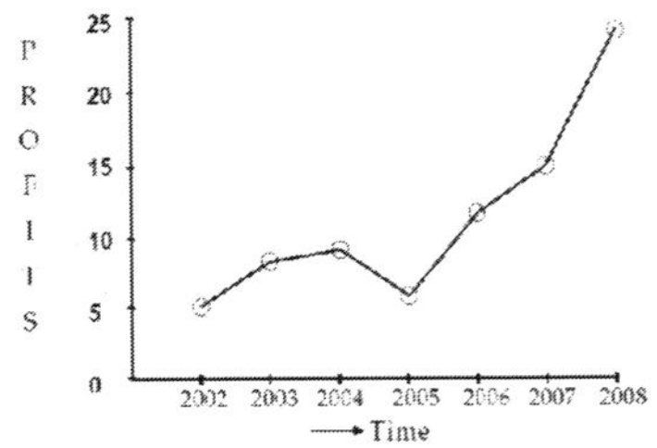

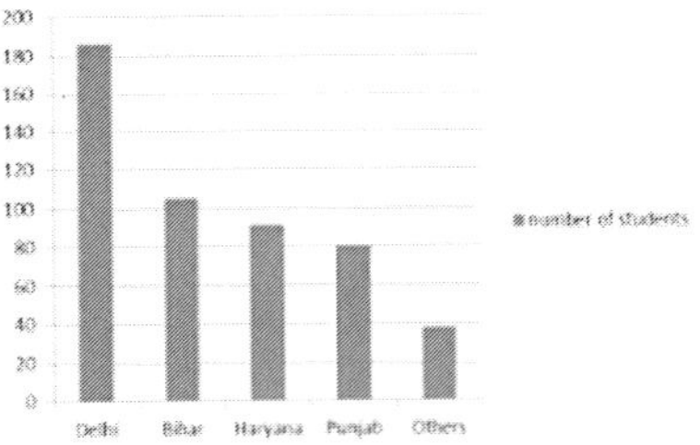

2. Circular/Pie diagram

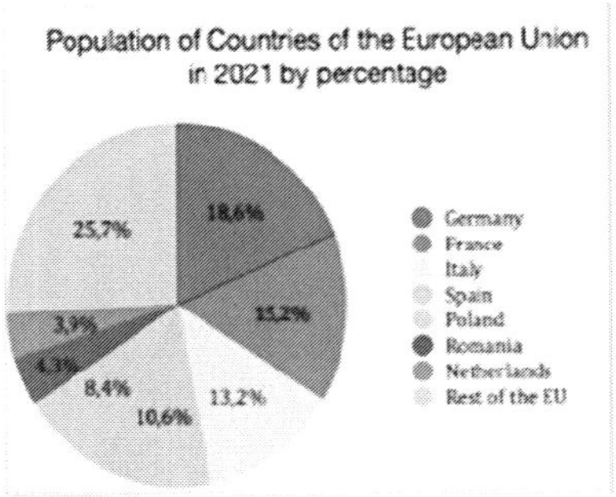

3. **Venn diagram**, is a widely used diagram style that shows the logical relation between sets, popularized by John Venn in the 1880s.

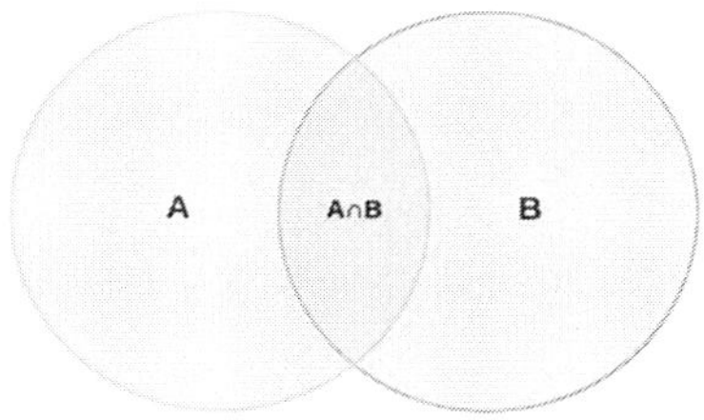

4. Two dimensional (or) Area Diagram

   Square diagram and rectangle diagrams

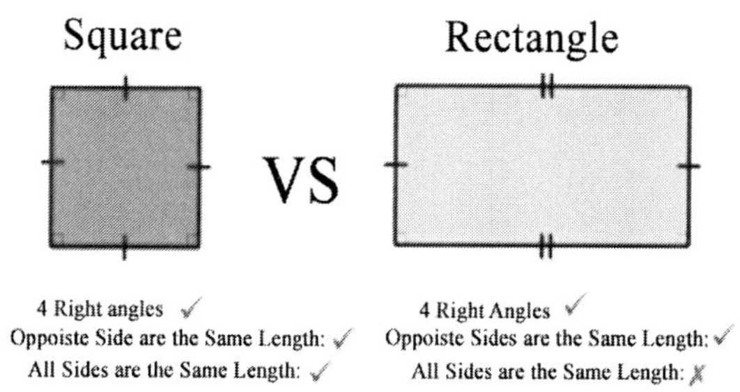

5. Three dimensional (or)Volume diagrams

   Cubes

   Spheres, Cylinders etc.

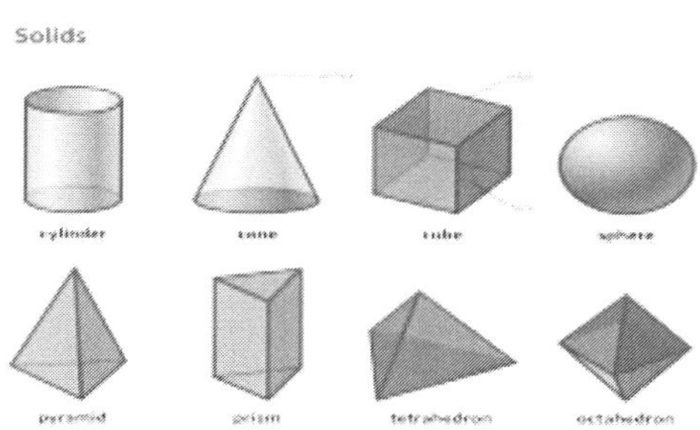

6. Pictogram

   Actual pictures

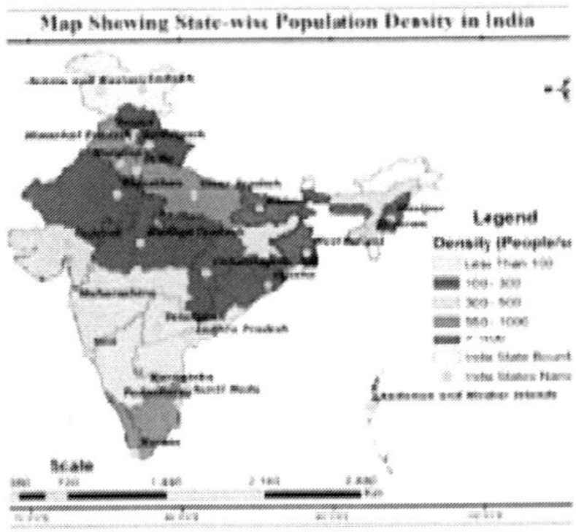

7. Cartogram: map and pictures

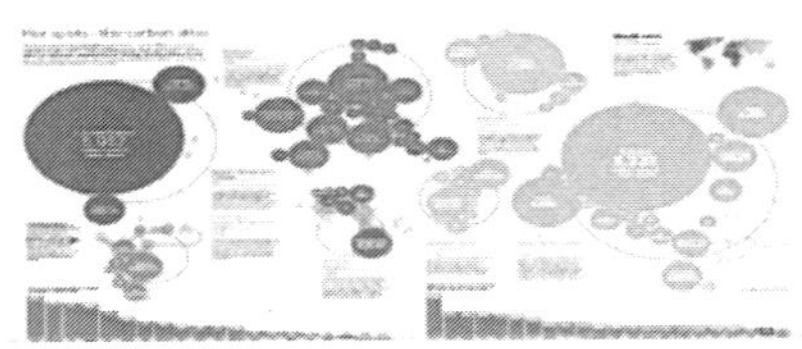

Answer

11. Enumerate merits and demerits of CRD and RBD

Advantages of a CRD/merits

I. Its layout is very easy.

II. There is complete flexibility in this design i.e. any number of treatments and replications for each treatment can be tried.

III. Whole experimental material can be utilized in this design (**homogenous set**).

IV. This design yields maximum degrees of freedom for experimental error.

V. The analysis of data is simplest as compared to any other design.

VI. Even if some **values are missing the analysis can be done**.

Disadvantages of a CRD/demerits

I. It is difficult to find homogeneous experimental units in all respects and hence CRD is seldom suitable for field experiments as compared to other experimental designs.

II. It is less accurate than other designs.

RBD is more efficient and accurate when compared to CRD (Advantages).

I. Chance of error in RBD is comparatively less.

II. Flexibility is also very high in RBD and thus any number of treatments and any number of replications can be used.

III. Statistical analysis is relatively simple and easy.

Disadvantages of RBD

I. When the number of treatments is increased, the block size will increase.

II. If the block size is large maintaining homogeneity is difficult and hence when more number of treatments is present this design may not be suitable.

III. RBD has problem with missing observations.

Long Questions

Answer

12. What do you mean by dispersion? What are various measures of dispersion? Describe about absolute mean deviation and coefficient of variation.

Dispersion is the state of getting dispersed or spread. Statistical dispersion means the extent to which a numerical data is likely to vary about an average value or mean. In other words, dispersion helps to understand the distribution of the data. E.g. Range, Quartile Deviation, Variance, Standard Deviation, Coefficient of variation, Standard Error, Mean Deviation.

Measure of Dispersion

a) Range

The **range** is defined as the difference between the largest and smallest observations. Range(X)= Max(X)- Min (X)

b) Quartile Deviation

The Quartile Deviation can be defined mathematically as half of the difference between the upper and lower quartile. Here, quartile deviation can be represented as QD; Q_3 denotes the upper quartile and Q_1 indicates the lower quartile. Quartile Deviation is also known as the Semi Interquartile range.

$$QD = \frac{1}{2}(Q_3 - Q_1)$$

c) Variance

The arithmetic mean of squared deviations is called the variance. It is defined as mean sum of the square deviation of the values when values are expressed as deviation from the population mean.

Variance formula

Population	**Sample**
$\sigma^2 = \frac{\Sigma(x_i - \mu)^2}{N}$	$s^2 = \frac{\Sigma(x_i - \overline{x})^2}{n-1}$
x_i = elements in population	x_i = elements in sample
μ = population mean	$\overline{x}$ = sample mean
N = population size	n = sample size

d) Standard Deviation (SD)

It shows how much variation there is from the average (mean). A low SD indicates that the data points tend to be close to the mean, whereas a high SD indicates that the data are spread out over a large range of values.

- Standard Deviation : + $\sqrt{Variance\ or\ \sigma}$

SD = + $\sqrt{[\Sigma x2 - (\Sigma x)2/N] / (N-1)}$.

e) Coefficient of variation

Sometimes the standard deviation is expressed as a percentage of the mean; we call this measure the coefficient of variation (CV). It is a dimensionless quantity that can be used for comparing **relative amounts of variation.** However, these comparisons are entirely subjective because its theoretical properties are complex, so we do not recommend its use.

$$CV = \frac{\sigma}{\mu} \times 100\%$$

σ= population standard deviation

μ= population mean

f) The standard error

The standard error of a statistic is the standard deviation of its sampling distribution or an estimate of that standard deviation. If the statistic is the sample mean, it is called the standard error of the mean.

$$SE = \frac{\sigma}{\sqrt{n}}$$

SE = standard error of the population

σ = population standard deviation

n = number of samples

The standard error of the mean, or simply standard error, indicates how different the population mean is likely to be from a sample mean.

g) Absolute mean deviation

What is Mean Absolute Deviation?

Mean Absolute Deviation (MAD) of a data set is the average distance between each data point of the data set and the mean of data. i.e. it represents the amount of variation that occurs around the mean value in the data set. It is also a measure of spread.

The Mean Absolute Deviation Formula for ungrouped data is given as follows:

$$MAD = \frac{\Sigma |x_i - \mu|}{n}$$

where, x_i represents the each observation of the dataset,

μ is the mean of the data set, and

n is the number of observations in the data set.

For group data,

$$MAD = \frac{\Sigma f_i |x_i - \overline{x}|}{\Sigma f_i}$$

Where,

x_i represents the each observation of the dataset,

bar{x} is mean of dataset

f_i represents frequency of corresponding observation x_i,

13. **Describe standard deviation with its merits and demerits.**

h) Standard Deviation (SD)

It shows how much variation there is from the average (mean). A low SD indicates that the data points tend to be close to the mean, whereas a high SD indicates that the data are spread out over a large range of values

Standard Deviation= + $\sqrt{Variance\ or\ \sigma}$

SD = + $\sqrt{[\Sigma x^2 - (\Sigma x)2/N] / (N-1)}$.

Some important points:

I. The concept of Standard Deviation was given by Karl Pearson in 1823.

II. Standard deviation is the **best measure of dispersion.**

III. It is based on each and every item of the distribution.

IV. We also know that the square root of any quantity is always positive or zero. Hence, the standard deviation is always non-negative.

Merits of SD

a. The value of standard deviation is based on every observation in a set of data.

b. It is less affected by fluctuations of sampling.

c. It is the only measure of variation capable of algebraic treatment.

d. Compared to other measures of dispersion, calculations of standard deviation is very easy.

Demerits of SD

a. It is affected very much by the extreme values of a series in as much as the squares of deviations of big items proportionately bigger than the squares of the smaller items.

b. It cannot be used for comparing the dispersion of two, or more series given in different units.

Answer

14. Differentiate between paired and unpaired 't' test with appropriate formula. 3 marks

Parameter of comparison	Paired t-test	Unpaired t-test
Meaning	Paired t-test, also known as repeated/related samples t-test, determines the distinction between the two means of the same samples.	Unpaired t-tests, also known as independent t-tests or student's t-test, is determining the two means groups of different/unrelated subjects.

Homogeneity of variances	Under Paired T-Test the variance of the two mean groups are not equal.	Under Unpaired T-Test the variance of the two mean groups are equal. For not equal variance different test statics is followed.
Effects/impacts	Paired T-tests deal with very minor errors since the test is done only between two similar groups.	Unpaired T-Tests have slightly more errors in comparison with paired T-Tests since the experimenter would be affected by variations between two different subjects.
Outcome	Paired T-Tests need not collect massive amounts of sample data for comparison this successively saves money and time.	Since Unpaired T-Tests have to compare the means of two independent subjects this lands up being a slightly more costlier and time-consuming process.
Test statistics	$t\frac{\bar{d}-0}{s_d/\sqrt{n}}$	$t=\frac{(\bar{y}_1-\bar{y}_2)-0}{\sqrt{\left(\frac{S_1^2}{n_1}+\frac{S_2^2}{n_2}\right)}}$

Q4. Write short notes.

15. What is Standard deviation? Give its computation formula.

1. Standard Deviation (SD)

It shows how much variation there is from the average (mean). A low SD indicates that the data points tend to be close to the mean, whereas a high SD indicates that the data are spread out over a large range of values.

Standard Deviation= + $\sqrt{\text{Variance}}$ *or*

SD = + $\sqrt{[\Sigma x^2 - (\Sigma x)2/N] / (N-1)}$.

Some important points

V. The concept of Standard Deviation was given by Karl Pearson in 1823.

VI. Standard deviation is the **best measure of dispersion.**

VII. It is based on each and every item of the distribution.

VIII. We also know that the square root of any quantity is always positive or zero. Hence, the standard deviation is always non-negative.

Merits of SD

e. The value of standard deviation is based on every observation in a set of data.

f. It is less affected by fluctuations of sampling.

g. It is the only measure of variation capable of algebraic treatment.

h. Compared to other measures of dispersion, calculations of standard deviation is very easy.

Demerits of SD

c. It is affected very much by the extreme values of a series in as much as the squares of deviations of big items proportionately bigger than the squares of the smaller items.

d. It cannot be used for comparing the dispersion of two, or more series given in different units.

16. What is the relationship between mean, median and mode?

In statistics, for a moderately skewed distribution, there exists a relation between mean, median, and mode. This **mean median and mode relation** is known as the "**empirical relationship**" which is defined as **Mode is equal to the difference between 3 times the Median and 2 times the Mean**.

Mathematically, it is written as Mean – Mode =3(Mean – Median).

The above **relation between mean median and mode** can be rewritten to find any of the measures of central tendencies when the remaining two are known.

- Formula to find Mean when Median and Mode are known: Mean =(3 Median – Mode)/2
- Formula to find Median when Mean and Mode are known: Median =(2 Mean + Mode)/3
- **Formula to find Mode when Mean and Median are known:**

  Mode=3 Median- 2 Mean

17. Define standard error

The standard error of a statistic is the standard deviation of its sampling distribution or an estimate of that standard deviation. If the statistic is the sample mean, it is called the standard error of the mean.

$$SE = \frac{\sigma}{\sqrt{n}}$$

SE = standard error of the population

σ = population standard deviation

n = number of samples

The standard error of the mean, or simply standard error, indicates how different the population mean is likely to be from a sample mean.

18. Define type I and type II error.

a. We have rejected the null hypothesis when it is true. We are making a **Type I error.** Is denoted by α, the probability of making a Type I error, generally in an experiment count at 5 or 1% level.

b. We have not rejected or we accepted the null hypothesis when it is false. We are making a **Type II error.** The probability of making a Type II error is usually designated by β (beta). Power of test is defined by 1- β.

19. Describe in brief

a. Describe Methods for collection of data

A statistical investigation always begins with collection of data. One can collect the data either by himself or from available records.

The data collected by the investigator himself or by his agent from the sample or population are called as the primary data. The source from which one gathers primary data is called as the **primary source**.

The data collected from the available sources is known as secondary data. The source from which we are getting secondary data is known as **secondary source**.

Data is collected by,

- Direct personal observation: The investigator himself goes to the field of enquiry and collects the data.
- Indirect personal observation: The investigator collects data from a third person (called as witness), who knows about the data being gathered.

b. Enlist the importance of frequency distribution in data handling.

It shows whether the observations are high or low and also whether they are concentrated in one area or spread out across the entire scale. Thus, frequency distribution presents a picture of how the individual observations are distributed in the measurement scale.

c. What is the test hypothesis? Define null and alternative hypothesis.

Hypothesis testing is a process that is concerned with making conclusions about the population using the information obtained from a sample.

There are two statistical hypotheses: the null hypothesis (H0) and the alternative hypothesis (H1). The null hypothesis is usually an assumption of unchanged state. For example, the H0 states that there is no difference between some characteristics, for example means or variances of two populations.

The alternative hypothesis, H1, describes a changed state or existence of a difference.

d. What are the measures of shape of distribution? Explain briefly about skewness of data.

The measures of the **shape of a distribution** are the coefficients of skewness and kurtosis.

***Skewness (sk)* is a measure of asymmetry of a frequency distribution**. It shows if deviations from the mean are larger on one side than the other side of the distribution. If the population mean (μ) is known, then skewness is:

$$sk = \frac{1}{(n-1)(n-2)} \sum_i \left(\frac{y_i - \mu}{s} \right)^3$$

For a symmetric distribution skewness is equal to zero. It is positive when the right tail is longer, and negative when left tail is longer.

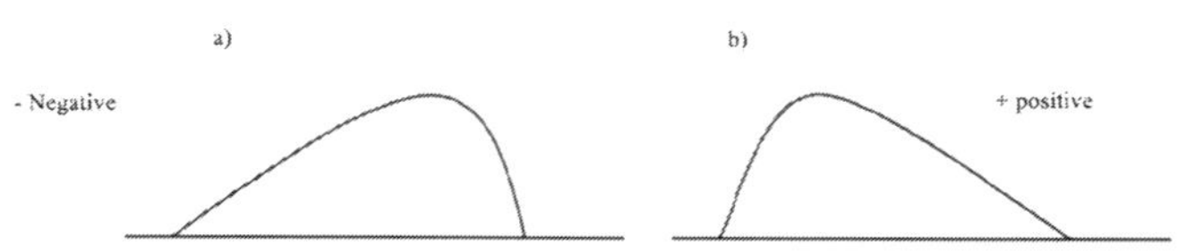

***Kurtosis (kt)* is a measure of flatness or steepness of a distribution**, or a measure of the **heaviness of the tails of a distribution**. If the population mean (μ) is known, kurtosis is:

$$kt = \frac{1}{2} \sum_i \left(\frac{y_i - \mu}{s} \right)^4 - 3$$

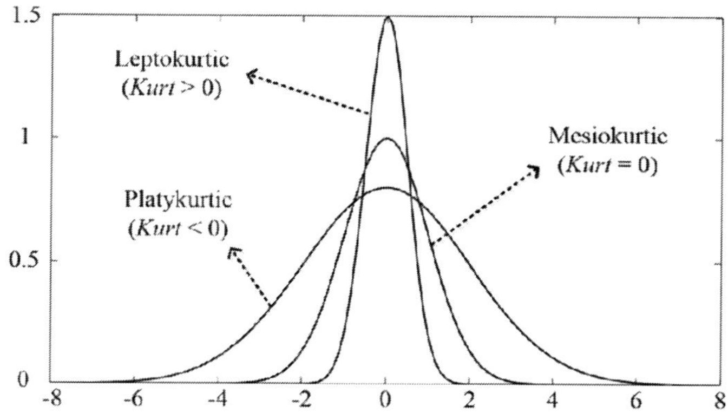

e. Enumerate merits and demerits of CRD and RBD.

Advantages of a CRD/merits

VII. Its layout is very easy.

VIII. There is complete flexibility in this design i.e. any number of treatments and replications for each treatment can be tried.

IX. Whole experimental material can be utilized in this design (homogenous set).

X. This design yields maximum degrees of freedom for experimental error.

XI. The analysis of data is simplest as compared to any other design.

XII. Even if some values are missing the analysis can be done.

Disadvantages of a CRD/demerits

III. It is difficult to find homogeneous experimental units in all respects and hence CRD is seldom suitable for field experiments as compared to other experimental designs.

IV. It is less accurate than other designs.

RBD is more efficient and accurate when compared to CRD (Advantages).

IV. Chance of error in RBD is comparatively less.

V. Flexibility is also very high in RBD and thus any number of treatments and any number of replications can be used.

VI. Statistical analysis is relatively simple and easy.

VII. Statistical analysis simple when one value is missing.

Disadvantages of RBD

IV. When the number of treatments is increased, the block size will increase.

V. If the block size is large maintaining homogeneity is difficult and hence when more number of treatments is present this design may not be suitable.

VI. RBD has problem with missing observations.

f) Define moments

In statistics, moments are measures of the shape and variability of a data set. They are used to describe the location and dispersion of the data. There are several types of moments that can be calculated, each providing different information about the data set.

- Arithmetic mean of various power expressed as deviation from mean in any distribution is called the moment of the distribution.
- **Central Moments:** the moment about the mean is called central moment. The rth moment of n observation about the mean X or central moment usually denoted by μr is define as.

  $\mu_r = \Sigma(X-\bar{X})^r/N$

  For raw data where X1, X2, X3........Xn are the 'n' observations and

  $\bar{X} = \Sigma(X)/n$

- The mean of first power of deviation is the first moment about the mean.

  $\mu_1 = \Sigma(X-\bar{X})^1/N$

- Since the sum of deviations of observation from arithmetic mean is always zero, the first moment μ1 is always zero.
- The second moment about the mean μ2 equal to **variance** as follow:

  $\mu_2 = \Sigma(X-\bar{X})^2/N$

- Third moment about the mean μ3 define as Skewness as follow:

  $\mu_3 = \Sigma(X-\bar{X})^3/N$

- The moments can be extended to higher power in the similar way but generally first four or up to μ4 define as **Kurtosis** is calculated:

  $\mu_4 = \Sigma(X-\bar{X})^4/N$

Answer

g) What is t-test? Who discovered it?

In 1908 William Sealy Gosset, an Englishman publishing under the pseudonym Student, developed the t-test and t distribution.

Hypothesis Test of a Population Mean for a Small Sample

The student t-distribution is used for testing hypotheses about the population mean for a small sample (say $n < 30$) drawn from a normal population. The test statistic is a t random variable:

$$\frac{\overline{}}{s \quad \sqrt{n}}$$

For a two-sided test the null hypothesis $H0$: $\mu = \mu 0$ is rejected if $|t| > t\alpha/2$, where $t\alpha/2$ is a critical value such that $P(t > t\alpha/2) = \alpha/2$.

Q4. Write short notes on following

20. Test of a Population Mean (Z test) versus t-test

One use of a hypothesis test is to determine if a sample mean is significantly different from a predetermined value. This example of hypothesis testing will be used to show the general principles of statistical hypotheses. First, a researcher must define null and alternative hypotheses. **To determine if a population mean is different from some value $\mu 0$**, the null and alternative hypotheses are:

$H0: \mu = \mu 0$

$H1: \mu \neq \mu 0$

The null hypothesis, $H0$, states that the population mean is equal to $\mu 0$, the alternative hypothesis, $H1$, states that the population mean is different from $\mu 0$.

The next step is to define an estimator of the population mean. This is the sample mean X. Now, a test statistic with a known theoretical distribution is defined. For large samples the sample mean has a normal distribution, so a standard normal variable is defined:

$$Z - test = \frac{\overline{x} - \mu}{\sigma / \sqrt{n}}$$

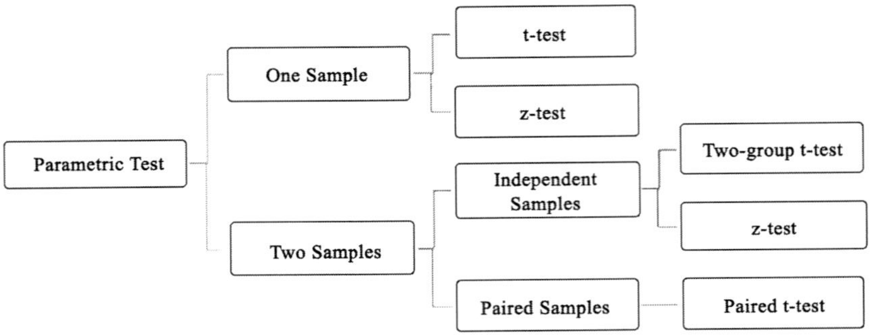

Basis for Comparison	**t-TEST**	**Z-TEST**
Meaning	t-test refers to a type of parametric test that is applied to identify, how the means of two sets of data differ from one another when variance is not given.	Z-test implies a hypothesis test which ascertains if the means of two datasets are different from each other when variance is given.
Based on	Student-t distribution	Normal distribution
Population variance	Unknown	Known
Sample Size	Small	Large
Test- statistics	$t - test = \frac{\overline{x} - \mu}{s / \sqrt{n}}$ x̄ is the sample mean s is sample standard deviation n is sample size μ is the population mean	$Z - test = \frac{\overline{x} - \mu}{\sigma / \sqrt{n}}$ x̄ is the sample mean σ is population standard deviation n is sample size μ is the population mean

21. Relationship between arithmetic, geometric and harmonic mean.

The arithmetic mean is a number that may be obtained by dividing the sum of the values in a set by the number of values in that set. If a1, a2, a3,…,an is a number of groups of values or the Arithmetic Progression, then the following statements are true:

$AM = (a_1+a_2+a_3+....,+a_n n)/n$

$GM = \sqrt[n]{(a_1.a_2.a_3.....,.a_n)}$

$HM = n/[(1/a_1) + (1/a_2) + (1/a_3) + ...+ (1/a_n)]$

Using the following phrase, we may depict the relationship between AM, GM, and HM:

AM > GM > HM

Arithmetic mean for two numbers a and b is the sum of the two numbers divided by two in the case of two numbers a and b.

AM = (a+b)/2

GM =√ab

HM = 2ab/a+b

AM × HM = GM²

Let us make an effort to comprehend this formula clearly, while also deriving this formula.

$$AM \times HM = \frac{a+b}{2} \times \frac{2ab}{a+b} = ab$$

$$ab = \sqrt{(ab)^2} = GM^2$$

22. Bayes' theorem

Bayes' theorem is named after the Reverend Thomas Bayes, also a statistician and philosopher. Bayes used conditional probability to provide an algorithm that uses evidence to calculate limits on an unknown parameter. His work was published in 1763.

Bayes theorem is useful for stating the probability of some event A, if there is information about the probability of some event E that happened after the event A.

Generally, there is an event A with k possible outcomes $A1, A2, \ldots, Ak$, that are independent and the sum of their probabilities is 1, ($\Sigma i\ P(Ai) = 1$). Also, there is an event E, that occurs after event A. Then:

$$P(A_i / E) = \frac{P(A_i \cap E)}{P(E)} = \frac{P(A_i)P(E / A_i)}{P(A_1)P(E / A_1) + P(E / A_2) + \ldots\ldots + P(A_k)P(E / A_k)}$$

23. Critical Difference (C.D.)

If F is significant or highly significant, critical difference between treatment means is to be worked out. **Critical difference** between any two treatment means is defined as the least significant difference between any two treatment means, to be exceeded by the difference between two treatment means to declare them as significantly different.

Critical difference between any two treatment means at 5% level = Standard error of the difference between the treatment means x table value of ‘t’ for error d.f. at 5% level.

Critical difference between any two treatment means at 1% level = Standard error of the difference between the treatment means x table value of ‘t’ for error d.f. at 1% level.

Critical difference between Tr1 and Tr2 at 5% (1%)

$$\sqrt{EMS\left(\frac{1}{n_1}+\frac{1}{n_2}\right)} \times 't'\, error\, d.f\; at\; 5\%\; (1\%)$$

Critical difference between Tr1,Tr3 at 5%(1%)

$$\sqrt{EMS\left(\frac{1}{n_1}+\frac{1}{n_3}\right)} \times 't'\, error\, d.f\; at\, 5\% (1\%)\, and\; so\, on...$$

24. What is binomial distribution?

Assume a single trial that can take only two outcomes, for example, Yes and No, success and failure, or 1 and 0. Such a variable is called a binary or Bernoulli variable. The probability distribution of a random variable y is determined by the parameter p and the number of trials n:

$$P(y) = \binom{n}{y} p^y q^{n-y} \quad y = 0,1,2,...,n$$

where:

p = the probability of success in a single trial

q = 1 -p = the probability of failure in a single trial

The expectation and variance of a binomial variable are:

E(y) = μ = np and Var(y) = σ^2 = npq

It can be expressed as

$P(y) = {}^nC_y \,.\, p^y (1 - p)^{n-y}$

25. What is poison distribution?

The Poisson distribution is a model for the relative frequency of rare events and data defined as counts and often is used for determination of the probability that some event will happen in a specific time, volume or area. For example, the number of microorganisms within a microscope field, or the number of mutations or distribution of animals from some plot may have a Poisson distribution.

A Poisson random variable y is defined as how many times some event occurs in specific time, or given volume or area. If we know that each single event occurs with the same probability, that is, the probability that some event will occur is equal for any part of time, volume or area, and the expected number of events is λ, then the probability function is defined as:

$$P(y) = \frac{e^{-\lambda}\lambda^{y}}{y!}$$

where λ is the average number of successes in a given time, volume or area, and e is the base of the natural logarithm ($e = 2.71828$).

Often, instead of the expected number, the proportion of successes is known, which is an estimate of the probability of success in a single trial (p). When p is small and the total number of trials (n) large, the binomial distribution can be approximated with a Poisson distribution, $\lambda = np$. A characteristic of the Poisson variable is that both the expectation and variance are equal to the parameter λ:

$E(y) = \mu = \lambda$ and $Var(y) = \sigma^2 = \lambda$ where, $\lambda = np$

26. What is chi-square distribution?

Consider a set of standard normal random variables zi, ($i = 1,\ldots, v$), that are identical and independently distributed with the mean $\mu = 0$ and standard deviation $\sigma = 1$. Define a random variable:

$\chi 2 = \Sigma i\, z^2 i\ i = 1,\ldots, v$

The variable $\chi 2$ has a *chi*-square distribution with v degrees of freedom. The shape of the *chi*-square distribution depends on degrees of freedom.

The expression on the right is of the from that **Karl Pearson** would generalize to the form

$$\chi^2 = \sum_{i=1}^{n} \frac{(O_i - E_i)^2}{E_i}$$

χ^2 = chi squared

Oi = observed value

Ei = expected value

27. Define rank correlation?

i) It measures the degree of association between the ranks of two variables.

ii) Concept given by Spearman. It is non parametric or qualitative test.

iii) No unit

iv) Ranges from -1 to + 1

v) $\rho = 1 - \frac{6\sum di^2}{n(n^2 - 1)}$ where ρ is lower case rho

Where,

di = xi – yi

xi = ith rank of x variable

yi = ith rank of y variable

n = pair number of observation

Merits

a. Very simple and easy to calculate.
b. This is useful for calculating correlation coefficient of between qualitative data.

Demerits

a. This method can use for only small number of observations.
b. It is very difficult for a large number of data set to calculate.

28. What are the different types of bioassays? Define LD50, ED50, LC50 and Therapeutic index.

A bioassay is an analytical method to determine the concentration or potency of a substance by its effect on living animals or plants (in vivo), or on living cells or in vitro

Bio assays are thus a type of experiments with the object of comparing the efficacy of two or more substances, or preparations, like drugs, by using responses produced by them on suitable living organisms.

Types of bioassay:

1. **Direct assay**, the stimulus applied to the subject is specific and directly measurable, and the response to that stimulus is recorded. (e.g. death of the subject).

2. **Indirect assay**, the stimulus is fixed in advance and the response is measured in the subjects.

   a. Quantitative response: The measurement of the response to the stimulus is on a continuous scale. Eg. ELISA plate reading.

   b. Quantal response: The response is binary; it is a determination of whether or not an event occurs (Yes or No, died or alive).

LD50-LD stands for "**Lethal Dose**". LD50is the amount of a material, given all at once, which causes the death of 50% (one half) of a group of test animals. The LD50is one way to measure the short-term poisoning potential (acute toxicity) of a material.

- LC50-LC stands for "**Lethal Concentration**".The concentrations of the chemical in air that kills 50% of the test animals during the observation period (traditional 4 hours)is the LC50value.
- ED50-ED stands for "**Effective Dose**". The ED50 is the dose of a medication that produces a desired pharmacologic effect in 50% of the studied patient population that takes the medication.

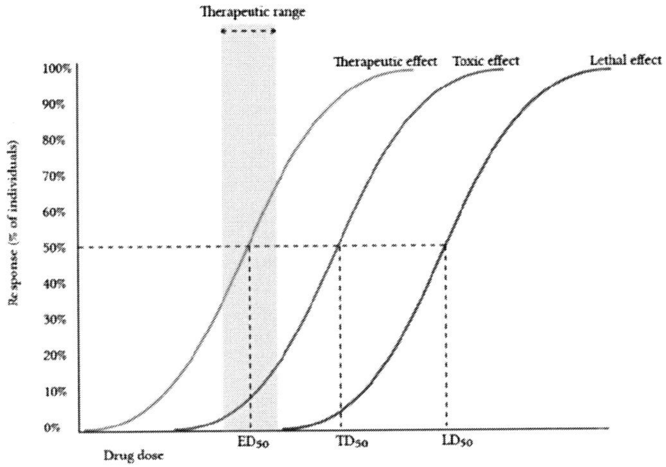

Therapeutic ratio/ Therapeutic index-It is comparison of amount of a therapeutic agent that causes the therapeutic effect to the amount that causes death.

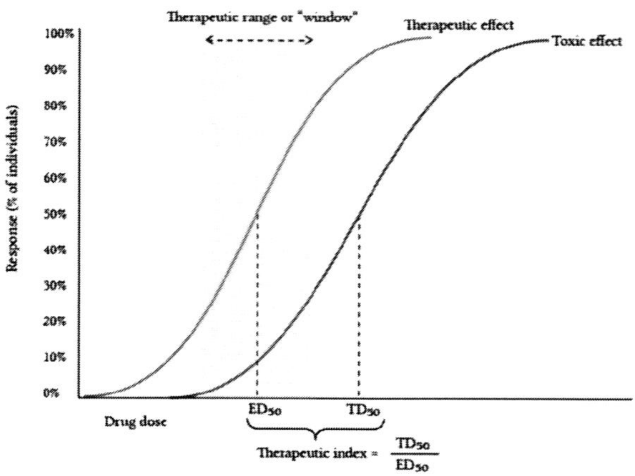

29. What is confidence interval?

Answer: The best way of establishing whether the estimate is good is to calculate what is called the **confidence interval for the mean**. This interval, defined by its upper and lower limits (the **confidence limits**), is generally interpreted as the range of values within which we expect the true population mean to lie with a certain probability.

Calculating the confidence interval for the mean

a) **Where σ is known:** Provided we have knowledge of σ, the 95% confidence interval for the mean is $\overline{X} \pm 1.96 \times$ SEM, i.e.

$$\overline{x} \pm 1.96\frac{\sigma}{\sqrt{n}} = \left(\overline{x} - 1.96\frac{\sigma}{\sqrt{n}}, \overline{x} + 1.96\frac{\sigma}{\sqrt{n}}\right)$$

Here, the upper and lower limits of the confidence interval within the bracket are separated

by the comma. For the 95% confidence interval, the multiple is 1.96 (often approximated by 2). The multiple is 2.58 for a 99% confidence interval; note that a multiple of 1.00 only gives approximately a 68% confidence interval.

(b) Where σ is unknown

Usually we do not know the value of the population standard deviation, σ, so we replace it by the sample estimate

$$\overline{x} \pm t_{0.05}\frac{s}{\sqrt{n}} = \left(\overline{x} - t_{0.05}\frac{s}{\sqrt{n}}, \overline{x} + t_{0.05}\frac{s}{\sqrt{n}}\right)$$

References

Biostatistical Analysis by Zar, Jerrold H., ISBN 9780131008465, 10 November 2009

Biostatistics for Animal Science by Miroslav Kaps and William R. Lamberson (CABI Publishing), 2004.

byjus.com web base information

Design and Analysis of Experiments by MN Das and NC Giri

Fundamentals of applied statistics – S. C. Gupta and V. K. Kapoor

Fundamentals of Biostatistical Analysis –Bernard Rosner, publication date 1995

Fundamentals of Statistics – S.C. Gupta

Handerson CR, Best Linear Unbiased Prediction Using Relationship Matrices Derived from Selected Base Populations. Journal of Dairy Science Vol. 68, No. 2, 1985

Henderson CR. Best linear unbiased estimation and prediction under a selection model. Biometrics. 1975 Jun;31(2):423-47.

http://www.fao.org/ag/againfo/resources/en/publications/sector_bri efs/lsb_IND.pdf

Instruction Manual for Integrated Sample Survey, provided by AHS Division, DADF, Ministry of Agriculture, GOI

Introductory Statistics, Prem S. Mann, Seventh Edition

Lourenco D, Legarra A, Tsuruta S, Masuda Y, Aguilar I, Misztal I. Single-Step Genomic Evaluations from Theory to Practice: Using SNP Chips and Sequence Data in BLUPF90. Genes (Basel). 2020 Jul 14;11(7):790. doi: 10.3390/genes11070790.

Nazarian A, Gezan SA. GenoMatrix: A Software Package for Pedigree-Based and Genomic Prediction Analyses on Complex Traits. J Hered. 2016 Jul;107(4):372-9.

Peter D. Cockcroft MA, VetMB, MSc, DCHP, DVM&S, MRCVS,, Mark A. Holmes PhD, MA, VetMB, MRCVS, Handbook of Evidence-based Veterinary Medicine, First published:28 March 2003, Print ISBN:9781405108904 |Online ISBN:9780470690833 |DOI:10.1002/9780470690833

Statistical Methods by SP Gupta

Statistical Methods for Agricultural Workers by VG Panse and PV Sukhatme

Statistical Methods for Biological workers – Pillai, S. K and Sinha, H. C

Statistical methods – George W. Snecdecor, William G. Cochran

Statistics for Veterinary and Animal Science, Aviva Petrie and Paul Watson, Third edition (Wiley and Blackwell)

wikipedia.org web base encyclopedia

Appendix

Critical values of chi-square (right tail)

Degrees of freedom (*df*)	Significance level (α) .99	.975	.95	.9	.1	.05	.025	.01
1	-------	0.001	0.004	0.016	2.706	3.841	5.024	6.635
2	0.020	0.051	0.103	0.211	4.605	5.991	7.378	9.210
3	0.115	0.216	0.352	0.584	6.251	7.815	9.348	11.345
4	0.297	0.484	0.711	1.064	7.779	9.488	11.143	13.277
5	0.554	0.831	1.145	1.610	9.236	11.070	12.833	15.086
6	0.872	1.237	1.635	2.204	10.645	12.592	14.449	16.812
7	1.239	1.690	2.167	2.833	12.017	14.067	16.013	18.475
8	1.646	2.180	2.733	3.490	13.362	15.507	17.535	20.090
9	2.088	2.700	3.325	4.168	14.684	16.919	19.023	21.666
10	2.558	3.247	3.940	4.865	15.987	18.307	20.483	23.209
11	3.053	3.816	4.575	5.578	17.275	19.675	21.920	24.725
12	3.571	4.404	5.226	6.304	18.549	21.026	23.337	26.217
13	4.107	5.009	5.892	7.042	19.812	22.362	24.736	27.688
14	4.660	5.629	6.571	7.790	21.064	23.685	26.119	29.141
15	5.229	6.262	7.261	8.547	22.307	24.996	27.488	30.578
16	5.812	6.908	7.962	9.312	23.542	26.296	28.845	32.000
17	6.408	7.564	8.672	10.085	24.769	27.587	30.191	33.409
18	7.015	8.231	9.390	10.865	25.989	28.869	31.526	34.805
19	7.633	8.907	10.117	11.651	27.204	30.144	32.852	36.191
20	8.260	9.591	10.851	12.443	28.412	31.410	34.170	37.566
21	8.897	10.283	11.591	13.240	29.615	32.671	35.479	38.932
22	9.542	10.982	12.338	14.041	30.813	33.924	36.781	40.289
23	10.196	11.689	13.091	14.848	32.007	35.172	38.076	41.638
24	10.856	12.401	13.848	15.659	33.196	36.415	39.364	42.980
25	11.524	13.120	14.611	16.473	34.382	37.652	40.646	44.314
26	12.198	13.844	15.379	17.292	35.563	38.885	41.923	45.642
27	12.879	14.573	16.151	18.114	36.741	40.113	43.195	46.963
28	13.565	15.308	16.928	18.939	37.916	41.337	44.461	48.278
29	14.256	16.047	17.708	19.768	39.087	42.557	45.722	49.588
30	14.953	16.791	18.493	20.599	40.256	43.773	46.979	50.892
40	22.164	24.433	26.509	29.051	51.805	55.758	59.342	63.691
50	29.707	32.357	34.764	37.689	63.167	67.505	71.420	76.154
60	37.485	40.482	43.188	46.459	74.397	79.082	83.298	88.379
70	45.442	48.758	51.739	55.329	85.527	90.531	95.023	100.425
80	53.540	57.153	60.391	64.278	96.578	101.879	106.629	112.329
100	61.754	65.647	69.126	73.291	107.565	113.145	118.136	124.116
1000	70.065	74.222	77.929	82.358	118.498	124.342	129.561	135.807

F Distribution Table

d.f.N. = degrees of freedom in numerator
d.f.D. = degrees of freedom in denominator

	$\alpha = 0.005$																		
	d.f.N.																		
d.f.D.	**1**	**2**	**3**	**4**	**5**	**6**	**7**	**8**	**9**	**10**	**12**	**15**	**20**	**24**	**30**	**40**	**60**	**120**	**∞**
1	16211	20000	21615	22500	23056	23437	23715	23925	24091	24224	24426	24630	24836	24940	25044	25148	25253	25359	25465
2	198.5	199.0	199.2	199.2	199.3	199.3	199.4	199.4	199.4	199.4	199.4	199.4	199.4	199.5	199.5	199.5	199.5	199.5	199.5
3	55.55	49.80	47.47	46.19	45.39	44.84	44.43	44.13	43.88	43.69	43.39	43.08	42.78	42.62	42.47	42.31	42.15	41.99	41.83
4	31.33	26.28	24.26	23.15	22.46	21.97	21.62	21.35	21.14	20.97	20.70	20.44	20.17	20.03	19.89	19.75	19.61	19.47	19.32
5	22.78	18.31	16.53	15.56	14.94	14.51	14.20	13.96	13.77	13.62	13.38	13.15	12.90	12.78	12.66	12.53	12.40	12.27	12.14
6	18.63	14.54	12.92	12.03	11.46	11.07	10.79	10.57	10.39	10.25	10.03	9.81	9.59	9.47	9.36	9.24	9.12	9.00	8.88
7	16.24	12.40	10.88	10.05	9.52	9.16	8.89	8.68	8.51	8.38	8.18	7.97	7.75	7.65	7.53	7.42	7.31	7.19	7.08
8	14.69	11.04	9.60	8.81	8.30	7.95	7.69	7.50	7.34	7.21	7.01	6.81	6.61	6.50	6.40	6.29	6.18	6.06	5.95
9	13.61	10.11	8.72	7.96	7.47	7.13	6.88	6.69	6.54	6.42	6.23	6.03	5.83	5.73	5.62	5.52	5.41	5.30	5.19
10	12.83	9.43	8.08	7.34	6.87	6.54	6.30	6.12	5.97	5.85	5.66	5.47	5.27	5.17	5.07	4.97	4.86	4.75	4.64
11	12.23	8.91	7.60	6.88	6.42	6.10	5.86	5.68	5.54	5.42	5.24	5.05	4.86	4.76	4.65	4.55	4.44	4.34	4.23
12	11.75	8.51	7.23	6.52	6.07	5.76	5.52	5.35	5.20	5.09	4.91	4.72	4.53	4.43	4.33	4.23	4.12	4.01	3.90
13	11.37	8.19	6.93	6.23	5.79	5.48	5.25	5.08	4.94	4.82	4.64	4.46	4.27	4.17	4.07	3.97	3.87	3.76	3.65
14	11.06	7.92	6.68	6.00	5.56	5.26	5.03	4.86	4.72	4.60	4.43	4.25	4.06	3.96	3.86	3.76	3.66	3.55	3.44
15	10.80	7.70	6.48	5.80	5.37	5.07	4.85	4.67	4.54	4.42	4.25	4.07	3.88	3.79	3.69	3.58	3.48	3.37	3.26
16	10.58	7.51	6.30	5.64	5.21	4.91	4.69	4.52	4.38	4.27	4.10	3.92	3.73	3.64	3.54	3.44	3.33	3.22	3.11
17	10.38	7.35	6.16	5.50	5.07	4.78	4.56	4.39	4.25	4.14	3.97	3.79	3.61	3.51	3.41	3.31	3.21	3.10	2.98
18	10.22	7.21	6.03	5.37	4.96	4.66	4.44	4.28	4.14	4.03	3.86	3.68	3.50	3.40	3.30	3.20	3.10	2.99	2.87
19	10.07	7.09	5.92	5.27	4.85	4.56	4.34	4.18	4.04	3.93	3.76	3.59	3.40	3.31	3.21	3.11	3.00	2.89	2.78
20	9.94	6.99	5.82	5.17	4.76	4.47	4.26	4.09	3.96	3.85	3.68	3.50	3.32	3.22	3.12	3.02	2.92	2.81	2.69
21	9.83	6.89	5.73	5.09	4.68	4.39	4.18	4.01	3.88	3.77	3.60	3.43	3.24	3.15	3.05	2.95	2.84	2.73	2.61
22	9.73	6.81	5.65	5.02	4.61	4.32	4.11	3.94	3.81	3.70	3.54	3.36	3.18	3.08	2.98	2.88	2.77	2.66	2.55
23	9.63	6.73	5.58	4.95	4.54	4.26	4.05	3.88	3.75	3.64	3.47	3.30	3.12	3.02	2.92	2.82	2.71	2.60	2.48
24	9.55	6.66	5.52	4.89	4.49	4.20	3.99	3.83	3.69	3.59	3.42	3.25	3.06	2.97	2.87	2.77	2.66	2.55	2.43
25	9.48	6.60	5.46	4.84	4.43	4.15	3.94	3.78	3.63	3.54	3.37	3.20	3.01	2.92	2.82	2.72	2.61	2.50	2.38
26	9.41	6.54	5.41	4.79	4.38	4.10	3.89	3.73	3.60	3.49	3.33	3.15	2.97	2.87	2.77	2.67	2.56	2.45	2.33
27	9.34	6.49	5.36	4.74	4.34	4.06	3.85	3.69	3.56	3.45	3.28	3.11	2.93	2.83	2.73	2.63	2.52	2.41	2.25
28	9.28	6.44	5.32	4.70	4.30	4.02	3.81	3.65	3.52	3.41	3.25	3.07	2.89	2.79	2.69	2.59	2.48	2.37	2.29
29	9.23	6.40	5.28	4.66	4.26	3.98	3.77	3.61	3.48	3.38	3.21	3.04	2.86	2.76	2.66	2.56	2.45	2.33	2.24
30	9.18	6.35	5.24	4.62	4.23	3.95	3.74	3.58	3.45	3.34	3.18	3.01	2.82	2.73	2.63	2.52	2.42	2.30	2.18
40	8.83	6.07	4.98	4.37	3.99	3.71	3.51	3.35	3.22	3.12	2.95	2.78	2.60	2.50	2.40	2.30	2.18	2.06	1.93
60	8.49	5.79	4.73	4.14	3.76	3.49	3.29	3.13	3.01	2.90	2.74	2.57	2.39	2.29	2.19	2.08	1.96	1.83	1.69
120	8.18	5.54	4.50	3.92	3.55	3.28	3.09	2.93	2.81	2.71	2.54	2.37	2.19	2.09	1.98	1.87	1.75	1.61	1.43
∞	7.88	5.30	4.28	3.72	3.35	3.09	2.90	2.74	2.62	2.52	2.36	2.19	2.00	1.90	1.79	1.67	1.53	1.36	1.00

F Distribution Table

d.f.D.	α = 0.01																		
	d.f.N.																		
	1	**2**	**3**	**4**	**5**	**6**	**7**	**8**	**9**	**10**	**12**	**15**	**20**	**24**	**30**	**40**	**60**	**120**	**∞**
1	4052	4999.5	5403	5625	5764	5859	5928	5982	6022	6056	6106	6157	6209	6235	6261	6287	6313	6339	6366
2	98.50	99.00	99.17	99.25	99.30	99.33	99.36	99.37	99.39	99.40	99.42	99.43	99.45	99.46	99.47	99.47	99,48	99.49	99.50
3	34.12	30.82	29.46	28.71	28.24	27.91	27.67	27.49	27.35	27.23	27.05	26.87	26.69	26.60	26.50	26.41	26.32	26.22	26.13
4	21.20	18.00	16.69	15.98	15.52	15.21	14.98	14.80	14.66	14.55	14.37	14.20	14.02	13.93	13.84	13.75	13.65	13.56	13.46
5	16.26	13.27	12.06	11.39	10.97	10.67	10.46	10.29	10.16	10.05	9.89	9.72	9.55	9.47	9.38	9.29	9.20	9.11	9.02
6	13.75	10.92	9.78	9.15	8.75	8.47	8.26	8.10	7.98	7.87	7.72	7.56	7.40	7.31	7.23	7.14	7.06	6.97	6.88
7	12.25	9.55	8.45	7.85	7.46	7.19	6.99	6.84	6.72	6.62	6.47	6.31	6.16	6.07	5.99	5.91	5.82	5.74	5.65
8	11.26	8.65	7.59	7.01	6.63	6.37	6.18	6.03	5.91	5.81	5.67	5.52	5.36	5.28	5.20	5.12	5.03	4.95	4.86
9	10.56	8.02	6.99	6.42	6.06	5.80	5.61	5.47	5.35	5.26	5.11	4.96	4.81	4.73	4.65	4.57	4.48	4.40	4.31
10	10.04	7.56	6.55	5.99	5.64	5.39	5.20	5.06	4.94	4.85	4.71	4.56	4.41	4.33	4.25	4.17	4.08	4.00	3.91
11	9.65	7.21	6.22	5.67	5.32	5.07	4.89	4.74	4.63	4.54	4.40	4.25	4.10	4.02	3.94	3.86	3.78	3.69	3.60
12	9.33	6.93	5.95	5.41	5.06	4.82	4.64	4.50	4.39	4.30	4.16	4.01	3.86	3.78	3.70	3.62	3.54	3.45	3.36
13	9.07	6.70	5.74	5.21	4.86	4.62	4.44	4.30	4.19	4.10	3.96	3.82	3.66	3.59	3.51	3.43	3.34	3.25	3.17
14	8.86	6.51	5.56	5.04	4.69	4.46	4.28	4.14	4.03	3.94	3.80	3.66	3.51	3.43	3.35	3.27	3.18	3.09	3.00
15	8.68	6.36	5.42	4.89	4.56	4.32	4.14	4.00	3.89	3.80	3.67	3.52	3.37	3.29	3.21	3.13	3.05	2.96	2.87
16	8.53	6.23	5.29	4.77	4.44	4.20	4.03	3.89	3.78	3.69	3.55	3.41	3.26	3.18	3.10	3.02	2.93	2.84	2.75
17	8.40	6.11	5.18	4.67	4.34	4.10	3.93	3.79	3.68	3.59	3.46	3.31	3.16	3.08	3.00	2.92	2.83	2.75	2.65
18	8.29	6.01	5.09	4.58	4.25	4.01	3.84	3.71	3.60	3.51	3.37	3.23	3.08	3.00	2.92	2.84	2.75	2.66	2.57
19	8.18	5.93	5.01	4.50	4.17	3.94	3.77	3.63	3.52	3.43	3.30	3.15	3.00	2.92	2.84	2.76	2.67	2.58	2.49
20	8.10	5.85	4.94	4.43	4.10	3.87	3.70	3.56	3.46	3.37	3.23	3.09	2.94	2.86	2.78	2.69	2.61	2.52	2.42
21	8.02	5.78	4.87	4.37	4.04	3.81	3.64	3.51	3.40	3.31	3.17	3.03	2.88	2.80	2.72	2.64	2.55	2.46	2.36
22	7.95	5.72	4.82	4.31	3.99	3.76	3.59	3.45	3.35	3.26	3.12	2.98	2.83	2.75	2.67	2.58	2.50	2.40	2.31
23	7.88	5.66	4.76	4.26	3.94	3.71	3.54	3.41	3.30	3.21	3.07	2.93	2.78	2.70	2.62	2.54	2.45	2.35	2.26
24	7.82	5.61	4.72	4.22	3.90	3.67	3.50	3.36	3.26	3.17	3.03	2.89	2.74	2.66	2.58	2.49	2.40	2.31	2.21
25	7.77	5.57	4.68	4.18	3.85	3.63	3.46	3.32	3.22	3.13	2.99	2.85	2.70	2.62	2.54	2.45	2.36	2.27	2.17
26	7.72	5.53	4.64	4.14	3.82	3.59	3.42	3.29	3.18	3.09	2.96	2.81	2.66	2.58	2.50	2.42	2.33	2.23	2.13
27	7.68	5.49	4.60	4.11	3.78	3.56	3.39	3.26	3.15	3.06	2.93	2.78	2.63	2.55	2.47	2.38	2.29	2.20	2.10
28	7.64	5.45	4.57	4.07	3.75	3.53	3.36	3.23	3.12	3.03	2.90	2.75	2.60	2.52	2.44	2.35	2.26	2.17	2.06
29	7.60	5.42	4.54	4.04	3.73	3.50	3.33	3.20	3.09	3.00	2.87	2.73	2.57	2.49	2.41	2.33	2.23	2.14	2.03
30	7.56	5.39	4.51	4.02	3.70	3.47	3.30	3.17	3.07	2.98	2.84	2.70	2.55	2.47	2.39	2.30	2.21	2.11	2.01
40	7.31	5.18	4.31	3.83	3.51	3.29	3.12	2.99	2.89	2.80	2.66	2.52	2.37	2.29	2.20	2.11	2.02	1.92	1.80
60	7.08	4.98	4.13	3.65	3.34	3.12	2.95	2.82	2.72	2.63	2.50	2.35	2.20	2.12	2.03	1.94	1.84	1.73	1.60
120	6.85	4.79	3.95	3.48	3.17	2.96	2.79	2.66	2.56	2.47	2.34	2.19	2.03	1.95	1.86	1.76	1.66	1.53	1.38
∞	6.63	4.61	3.78	3.32	3.02	2.80	2.64	2.51	2.41	2.32	2.18	2.04	1.88	1.79	1.70	1.59	1.47	1.32	1.00

F Distribution Table

d.f.D.	α = 0.025 d.f.N.																		
	1	2	3	4	5	6	7	8	9	10	12	15	20	24	30	40	60	120	∞
1	647.8	799.5	864.2	899.6	921.8	937.1	948.2	956.7	963.3	968.6	976.7	984.9	993.1	997.2	1001	1006	1010	1014	1018
2	38.51	39.00	39.17	39.25	39.30	39.33	39.36	39.37	39.39	39.40	39.41	39.43	39.45	39.46	39.46	39.47	39.48	39.49	39.50
3	17.44	16.04	15.44	15.10	14.88	14.73	14.62	14.54	14.47	14.42	14.34	14.25	14.17	14.12	14.08	14.04	13.99	13.95	13.90
4	12.22	10.65	9.98	9.60	9.36	9.20	9.07	8.98	8.90	8.84	8.75	8.66	8.56	8.51	8.46	8.41	8.36	8.31	8.26
5	10.01	8.43	7.76	7.39	7.15	6.98	6.85	6.76	6.68	6.62	6.52	6.43	6.33	6.28	6.23	6.18	6.12	6.07	6.02
6	8.81	7.26	6.60	6.23	5.99	5.82	5.70	5.60	5.52	5.46	5.37	5.27	5.17	5.12	5.07	5.01	4.96	4.90	4.85
7	8.07	6.54	5.89	5.52	5.29	5.12	4.99	4.90	4.82	4.76	4.67	4.57	4.47	4.42	4.36	4.31	4.25	4.20	4.14
8	7.57	6.06	5.42	5.05	4.82	4.65	4.53	4.43	4.36	4.30	4.20	4.10	4.00	3.95	3.89	3.84	3.78	3.73	3.67
9	7.21	5.71	5.08	4.72	4.48	4.32	4.20	4.10	4.03	3.96	3.87	3.77	3.67	3.61	3.56	3.51	3.45	3.39	3.33
10	6.94	5.46	4.83	4.47	4.24	4.07	3.95	3.85	3.78	3.72	3.62	3.52	3.42	3.37	3.31	3.26	3.20	3.14	3.08
11	6.72	5.26	4.63	4.28	4.04	3.88	3.76	3.66	3.59	3.53	3.43	3.33	3.23	3.17	3.12	3.06	3.00	2.94	2.88
12	6.55	5.10	4.47	4.12	3.89	3.73	3.61	3.51	3.44	3.37	3.28	3.18	3.07	3.02	2.96	2.91	2.85	2.79	2.72
13	6.41	4.97	4.35	4.00	3.77	3.60	3.48	3.39	3.31	3.25	3.15	3.05	2.95	2.89	2.84	2.78	2.72	2.66	2.60
14	6.30	4.86	4.24	3.89	3.66	3.50	3.38	3.29	3.21	3.15	3.05	2.95	2.84	2.79	2.73	2.67	2.61	2.55	2.49
15	6.20	4.77	4.15	3.80	3.58	3.41	3.29	3.20	3.12	3.06	2.96	2.86	2.76	2.70	2.64	2.59	2.52	2.46	2.40
16	6.12	4.69	4.08	3.73	3.50	3.34	3.22	3.12	3.05	2.99	2.89	2.79	2.68	2.63	2.57	2.51	2.45	2.38	2.32
17	6.04	4.62	4.01	3.66	3.44	3.28	3.16	3.06	2.98	2.92	2.82	2.72	2.62	2.56	2.50	2.44	2.38	2.32	2.25
18	5.98	4.56	3.95	3.61	3.38	3.22	3.10	3.01	2.93	2.87	2.77	2.67	2.56	2.50	2.44	2.38	2.32	2.26	2.19
19	5.92	4.51	3.90	3.56	3.33	3.17	3.05	2.96	2.88	2.82	2.72	2.62	2.51	2.45	2.39	2.33	2.27	2.20	2.13
20	5.87	4.46	3.86	3.51	3.29	3.13	3.01	2.91	2.84	2.77	2.68	2.57	2.46	2.41	2.35	2.29	2.22	2.16	2.09
21	5.83	4.42	3.82	3.48	3.25	3.09	2.97	2.87	2.80	2.73	2.64	2.53	2.42	2.37	2.31	2.25	2.18	2.11	2.04
22	5.79	4.38	3.78	3.44	3.22	3.05	2.93	2.84	2.76	2.70	2.60	2.50	2.39	2.33	2.27	2.21	2.14	2.08	2.00
23	5.75	4.35	3.75	3.41	3.18	3.02	2.90	2.81	2.73	2.67	2.57	2.47	2.36	2.30	2.24	2.18	2.11	2.04	1.97
24	5.72	4.32	3.72	3.38	3.15	2.99	2.87	2.78	2.70	2.64	2.54	2.44	2.33	2.27	2.21	2.15	2.08	2.01	1.94
25	5.69	4.29	3.69	3.35	3.13	2.97	2.85	2.75	2.68	2.61	2.51	2.41	2.30	2.24	2.18	2.12	2.05	1.98	1.91
26	5.66	4.27	3.67	3.33	3.10	2.94	2.82	2.73	2.65	2.59	2.49	2.39	2.28	2.22	2.16	2.09	2.03	1.95	1.88
27	5.63	4.24	3.65	3.31	3.08	2.92	2.80	2.71	2.63	2.57	2.47	2.36	2.25	2.19	2.13	2.07	2.00	1.93	1.85
28	5.61	4.22	3.63	3.29	3.06	2.90	2.78	2.69	2.61	2.55	2.45	2.34	2.23	2.17	2.11	2.05	1.98	1.91	1.83
29	5.59	4.20	3.61	3.27	3.04	2.88	2.76	2.67	2.59	2.53	2.43	2.32	2.21	2.15	2.09	2.03	1.96	1.89	1.81
30	5.57	4.18	3.69	3.25	3.03	2.87	2.75	2.65	2.57	2.51	2.41	2.31	2.20	2.14	2.07	2.01	1.94	1.87	1.79
40	5.42	4.05	3.46	3.13	2.90	2.74	2.62	2.53	2.45	2.39	2.29	2.18	2.07	2.01	1.94	1.88	1.80	1.72	1.64
60	5.29	3.93	3.34	3.01	2.79	2.63	2.51	2.41	2.33	2.27	2.17	2.06	1.94	1.88	1.82	1.74	1.67	1.58	1.48
120	5.15	3.80	3.23	2.89	2.67	2.52	2.39	2.30	2.22	2.16	2.05	1.94	1.82	1.76	1.69	1.61	1.53	1.43	1.31
∞	5.02	3.69	3.12	2.79	2.57	2.41	2.29	2.19	2.11	2.05	1.94	1.83	1.71	1.64	1.57	1.48	1.39	1.27	1.00

F Distribution Table

d.f.D.	α = 0.05 d.f.N. 1	2	3	4	5	6	7	8	9	10	12	15	20	24	30	40	60	120	∞
1	161.4	199.5	215.7	224.6	230.2	234.0	236.8	238.9	240.5	241.9	243.9	245.9	248.0	249.1	250.1	251.1	252.2	253.3	254.3
2	18.51	19.00	19.16	19.25	19.30	19.33	19.35	19.37	19.38	19.40	19.41	19.43	19.45	19.45	19.46	19.47	19.48	19.49	19.50
3	10.13	9.55	9.28	9.12	9.01	8.94	8.89	8.85	8.81	8.79	8.74	8.70	8.66	8.64	8.62	8.59	8.57	8.55	8.53
4	7.71	6.94	6.59	6.39	6.26	6.16	6.09	6.04	6.00	5.96	5.91	5.86	5.80	5.77	5.75	5.72	5.69	5.66	5.63
5	6.61	5.79	5.41	5.19	5.05	4.95	4.88	4.82	4.77	4.74	4.68	4.62	4.56	4.53	4.50	4.46	4.43	4.40	4.36
6	5.99	5.14	4.76	4.53	4.39	4.28	4.21	4.15	4.10	4.06	4.00	3.94	3.87	3.84	3.81	3.77	3.74	3.70	3.67
7	5.59	4.74	4.35	4.12	3.97	3.87	3.79	3.73	3.68	3.64	3.57	3.51	3.44	3.41	3.38	3.34	3.30	3.27	3.23
8	5.32	4.46	4.07	3.84	3.69	3.58	3.50	3.44	3.39	3.35	3.28	3.22	3.15	3.12	3.08	3.04	3.01	2.97	2.93
9	5.12	4.26	3.86	3.63	3.48	3.37	3.29	3.23	3.18	3.14	3.07	3.01	2.94	2.90	2.86	2.83	2.79	2.75	2.71
10	4.96	4.10	3.71	3.48	3.33	3.22	3.14	3.07	3.02	2.98	2.91	2.85	2.77	2.74	2.70	2.66	2.62	2.58	2.54
11	4.84	3.98	3.59	3.36	3.20	3.09	3.01	2.95	2.90	2.85	2.79	2.72	2.65	2.61	2.57	2.53	2.49	2.45	2.40
12	4.75	3.89	3.49	3.26	3.11	3.00	2.91	2.85	2.80	2.75	2.69	2.62	2.54	2.51	2.47	2.43	2.38	2.34	2.30
13	4.67	3.81	3.41	3.18	3.03	2.92	2.83	2.77	2.71	2.67	2.60	2.53	2.46	2.42	2.38	2.34	2.30	2.25	2.21
14	4.60	3.74	3.34	3.11	2.96	2.85	2.76	2.70	2.65	2.60	2.53	2.46	2.39	2.35	2.31	2.27	2.22	2.18	2.13
15	4.54	3.68	3.29	3.06	2.90	2.79	2.71	2.64	2.59	2.54	2.48	2.40	2.33	2.29	2.25	2.20	2.16	2.11	2.07
16	4.49	3.63	3.24	3.01	2.85	2.74	2.66	2.59	2.54	2.49	2.42	2.35	2.28	2.24	2.19	2.15	2.11	2.06	2.01
17	4.45	3.59	3.20	2.96	2.81	2.70	2.61	2.55	2.49	2.45	2.38	2.31	2.23	2.19	2.15	2.10	2.06	2.01	1.96
18	4.41	3.55	3.16	2.93	2.77	2.66	2.58	2.51	2.46	2.41	2.34	2.27	2.19	2.15	2.11	2.06	2.02	1.97	1.92
19	4.38	3.52	3.13	2.90	2.74	2.63	2.54	2.48	2.42	2.38	2.31	2.23	2.16	2.11	2.07	2.03	1.98	1.93	1.88
20	4.35	3.49	3.10	2.87	2.71	2.60	2.51	2.45	2.39	2.35	2.28	2.20	2.12	2.08	2.04	1.99	1.95	1.90	1.84
21	4.32	3.47	3.07	2.84	2.68	2.57	2.49	2.42	2.37	2.32	2.25	2.18	2.10	2.05	2.01	1.96	1.92	1.87	1.81
22	4.30	3.44	3.05	2.82	2.66	2.55	2.46	2.40	2.34	2.30	2.23	2.15	2.07	2.03	1.98	1.94	1.89	1.84	1.78
23	4.28	3.42	3.03	2.80	2.64	2.53	2.44	2.37	2.32	2.27	2.20	2.13	2.05	2.01	1.96	1.91	1.86	1.81	1.76
24	4.26	3.40	3.01	2.78	2.62	2.51	2.42	2.36	2.30	2.25	2.18	2.11	2.03	1.98	1.94	1.89	1.84	1.79	1.73
25	4.24	3.39	2.99	2.76	2.60	2.49	2.40	2.34	2.28	2.24	2.16	2.09	2.01	1.96	1.92	1.87	1.82	1.77	1.71
26	4.23	3.37	2.98	2.74	2.59	2.47	2.39	2.32	2.27	2.22	2.15	2.07	1.99	1.95	1.90	1.85	1.80	1.75	1.69
27	4.21	3.35	2.96	2.73	2.57	2.46	2.37	2.31	2.25	2.20	2.13	2.06	1.97	1.93	1.88	1.84	1.79	1.73	1.67
28	4.20	3.34	2.95	2.71	2.56	2.45	2.36	2.29	2.24	2.19	2.12	2.04	1.96	1.91	1.87	1.82	1.77	1.71	1.65
29	4.18	3.33	2.93	2.70	2.55	2.43	2.35	2.28	2.22	2.18	2.10	2.03	1.94	1.90	1.85	1.81	1.75	1.70	1.64
30	4.17	3.32	2.92	2.69	2.53	2.42	2.33	2.27	2.21	2.16	2.09	2.01	1.93	1.89	1.84	1.79	1.74	1.68	1.62
40	4.08	3.23	2.84	2.61	2.45	2.34	2.25	2.18	2.12	2.08	2.00	1.92	1.84	1.79	1.74	1.69	1.64	1.58	1.51
60	4.00	3.15	2.76	2.52	2.37	2.25	2.17	2.10	2.04	1.99	1.92	1.84	1.75	1.70	1.65	1.59	1.53	1.47	1.39
120	3.92	3.07	2.68	2.45	2.29	2.17	2.09	2.02	1.96	1.91	1.83	1.75	1.66	1.61	1.55	1.50	1.43	1.35	1.25
∞	3.84	3.00	2.60	2.37	2.21	2.10	2.01	1.94	1.88	1.83	1.75	1.67	1.57	1.52	1.46	1.39	1.32	1.22	1.00

F Distribution Table

	α = 0.10																		
	d.f.N.																		
d.f.D.	**1**	**2**	**3**	**4**	**5**	**6**	**7**	**8**	**9**	**10**	**12**	**15**	**20**	**24**	**30**	**40**	**60**	**120**	**∞**
1	39.86	49.50	53.59	55.83	57.24	58.20	58.91	59.44	59.86	60.19	60.71	61.22	61.74	62.00	62.26	62.53	62.79	63.06	63.33
2	8.53	9.00	9.16	9.24	9.29	9.33	9.35	9.37	9.38	9.39	9.41	9.42	9.44	9.45	9.46	9.47	9.47	9.48	9.49
3	5.54	5.46	5.39	5.34	5.31	5.28	5.27	5.25	5.24	5.23	5.22	5.20	5.18	5.18	5.17	5.16	5.15	5.14	5.13
4	4.54	4.32	4.19	4.11	4.05	4.01	3.98	3.95	3.94	3.92	3.90	3.87	3.84	3.83	3.82	3.80	3.79	3.78	3.76
5	4.06	3.78	3.62	3.52	3.45	3.40	3.37	3.34	3.32	3.30	3.27	3.24	3.21	3.19	3.17	3.16	3.14	3.12	3.10
6	3.78	3.46	3.29	3.18	3.11	3.05	3.01	2.98	2.96	2.94	2.90	2.87	2.84	2.82	2.80	2.78	2.76	2.74	2.72
7	3.59	3.26	3.07	2.96	2.88	2.83	2.78	2.75	2.72	2.70	2.67	2.63	2.59	2.58	2.56	2.54	2.51	2.49	2.47
8	3.46	3.11	2.92	2.81	2.73	2.67	2.62	2.59	2.56	2.54	2.50	2.46	2.42	2.40	2.38	2.36	2.34	2.32	2.29
9	3.36	3.01	2.81	2.69	2.61	2.55	2.51	2.47	2.44	2.42	2.38	2.34	2.30	2.28	2.25	2.23	2.21	2.18	2.16
10	3.29	2.92	2.73	2.61	2.52	2.46	2.41	2.38	2.35	2.32	2.28	2.24	2.20	2.18	2.16	2.13	2.11	2.08	2.06
11	3.23	2.86	2.66	2.54	2.45	2.39	2.34	2.30	2.27	2.25	2.21	2.17	2.12	2.10	2.08	2.05	2.03	2.00	1.97
12	3.18	2.81	2.61	2.48	2.39	2.33	2.28	2.24	2.21	2.19	2.15	2.10	2.06	2.04	2.01	1.99	1.96	1.93	1.90
13	3.14	2.76	2.56	2.43	2.35	2.28	2.23	2.20	2.16	2.14	2.10	2.05	2.01	1.98	1.96	1.93	1.90	1.88	1.85
14	3.10	2.73	2.52	2.39	2.31	2.24	2.19	2.15	2.12	2.10	2.05	2.01	1.96	1.94	1.91	1.89	1.86	1.83	1.80
15	3.07	2.70	2.49	2.36	2.27	2.21	2.16	2.12	2.09	2.06	2.02	1.97	1.92	1.90	1.87	1.85	1.82	1.79	1.76
16	3.05	2.67	2.46	2.33	2.24	2.18	2.13	2.09	2.06	2.03	1.99	1.94	1.89	1.87	1.84	1.81	1.78	1.75	1.72
17	3.03	2.64	2.44	2.31	2.22	2.15	2.10	2.06	2.03	2.00	1.96	1.91	1.86	1.84	1.81	1.78	1.75	1.72	1.69
18	3.01	2.62	2.42	2.29	2.20	2.13	2.08	2.04	2.00	1.98	1.93	1.89	1.84	1.81	1.78	1.75	1.72	1.69	1.66
19	2.99	2.61	2.40	2.27	2.18	2.11	2.06	2.02	1.98	1.96	1.91	1.86	1.81	1.79	1.76	1.73	1.70	1.67	1.63
20	2.97	2.59	2.38	2.25	2.16	2.09	2.04	2.00	1.96	1.94	1.89	1.84	1.79	1.77	1.74	1.71	1.68	1.64	1.61
21	2.96	2.57	2.36	2.23	2.14	2.08	2.02	1.98	1.95	1.92	1.87	1.83	1.78	1.75	1.72	1.69	1.66	1.62	1.59
22	2.95	2.56	2.35	2.22	2.13	2.06	2.01	1.97	1.93	1.90	1.86	1.81	1.76	1.73	1.70	1.67	1.64	1.60	1.57
23	2.94	2.55	2.34	2.21	2.11	2.05	1.99	1.95	1.92	1.89	1.84	1.80	1.74	1.72	1.69	1.66	1.62	1.59	1.55
24	2.93	2.54	2.33	2.19	2.10	2.04	1.98	1.94	1.91	1.88	1.83	1.78	1.73	1.70	1.67	1.64	1.61	1.57	1.53
25	2.92	2.53	2.32	2.18	2.09	2.02	1.97	1.93	1.89	1.87	1.82	1.77	1.72	1.69	1.66	1.63	1.59	1.56	1.52
26	2.91	2.52	2.31	2.17	2.08	2.01	1.96	1.92	1.88	1.86	1.81	1.76	1.71	1.68	1.65	1.61	1.58	1.54	1.50
27	2.90	2.51	2.30	2.17	2.07	2.00	1.95	1.91	1.87	1.85	1.80	1.75	1.70	1.67	1.64	1.60	1.57	1.53	1.49
28	2.89	2.50	2.29	2.16	2.06	2.00	1.94	1.90	1.87	1.84	1.79	1.74	1.69	1.66	1.63	1.59	1.56	1.52	1.48
29	2.89	2.50	2.28	2.15	2.06	1.99	1.93	1.89	1.86	1.83	1.78	1.73	1.68	1.65	1.62	1.58	1.55	1.51	1.47
30	2.88	2.49	2.28	2.14	2.05	1.98	1.93	1.88	1.85	1.82	1.77	1.72	1.67	1.64	1.61	1.57	1.54	1.50	1.46
40	2.84	2.44	2.23	2.09	2.00	1.93	1.87	1.83	1.79	1.76	1.71	1.66	1.61	1.57	1.54	1.51	1.47	1.42	1.38
60	2.79	2.39	2.18	2.04	1.95	1.87	1.82	1.77	1.74	1.71	1.66	1.60	1.54	1.51	1.48	1.44	1.40	1.35	1.29
120	2.75	2.35	2.13	1.99	1.90	1.82	1.77	1.72	1.68	1.65	1.60	1.55	1.48	1.45	1.41	1.37	1.32	1.26	1.19
∞	2.71	2.30	2.08	1.94	1.85	1.77	1.72	1.67	1.63	1.60	1.55	1.49	1.42	1.38	1.34	1.30	1.24	1.17	1.00

Critical values of t for one-tailed test
significance level (α)

Degrees of freedom *(df)*	0.2	0.15	0.1	0.05	0.025	0.01	0.005	0.001	0.0005
1	1.376	1.963	3.078	6.314	12.706	31.821	63.657	318.309	636.619
2	1.061	1.386	1.886	2.920	4.303	6.965	9.925	22.327	31.599
3	0.978	1.250	1.638	2.353	3.182	4.541	5.841	10.215	12.924
4	0.941	1.190	1.533	2.132	2.776	3.747	4.604	7.173	8.610
5	0.920	1.156	1.476	2.015	2.571	3.365	4.032	5.893	6.869
6	0.906	1.134	1.440	1.943	2.447	3.143	3.707	5.208	5.959
7	0.896	1.119	1.415	1.895	2.365	2.998	3.499	4.785	5.408
8	0.889	1.108	1.397	1.860	2.306	2.896	3.355	4.501	5.041
9	0.883	1.100	1.383	1.833	2.262	2.821	3.250	4.297	4.781
10	0.879	1.093	1.372	1.812	2.228	2.764	3.169	4.144	4.587
11	0.876	1.088	1.363	1.796	2.201	2.718	3.106	4.025	4.437
12	0.873	1.083	1.356	1.782	2.179	2.681	3.055	3.930	4.318
13	0.870	1.079	1.350	1.771	2.160	2.650	3.012	3.852	4.221
14	0.868	1.076	1.345	1.761	2.145	2.624	2.977	3.787	4.140
15	0.866	1.074	1.341	1.753	2.131	2.602	2.947	3.733	4.073
16	0.865	1.071	1.337	1.746	2.120	2.583	2.921	3.686	4.015
17	0.863	1.069	1.333	1.740	2.110	2.567	2.898	3.646	3.965
18	0.862	1.067	1.330	1.734	2.101	2.552	2.878	3.610	3.922
19	0.861	1.066	1.328	1.729	2.093	2.539	2.861	3.579	3.883
20	0.860	1.064	1.325	1.725	2.086	2.528	2.845	3.552	3.850
21	0.859	1.063	1.323	1.721	2.080	2.518	2.831	3.527	3.819
22	0.858	1.061	1.321	1.717	2.074	2.508	2.819	3.505	3.792
23	0.858	1.060	1.319	1.714	2.069	2.500	2.807	3.485	3.768
24	0.857	1.059	1.318	1.711	2.064	2.492	2.797	3.467	3.745
25	0.856	1.058	1.316	1.708	2.060	2.485	2.787	3.450	3.725
26	0.856	1.058	1.315	1.706	2.056	2.479	2.779	3.435	3.707
27	0.855	1.057	1.314	1.703	2.052	2.473	2.771	3.421	3.690
28	0.855	1.056	1.313	1.701	2.048	2.467	2.763	3.408	3.674
29	0.854	1.055	1.311	1.699	2.045	2.462	2.756	3.396	3.659
30	0.854	1.055	1.310	1.697	2.042	2.457	2.750	3.385	3.646
40	0.851	1.050	1.303	1.684	2.021	2.423	2.704	3.307	3.551
50	0.849	1.047	1.299	1.676	2.009	2.403	2.678	3.261	3.496
60	0.848	1.045	1.296	1.671	2.000	2.390	2.660	3.232	3.460
70	0.847	1.044	1.294	1.667	1.994	2.381	2.648	3.211	3.435
80	0.846	1.043	1.292	1.664	1.990	2.374	2.639	3.195	3.416
100	0.845	1.042	1.290	1.660	1.984	2.364	2.626	3.174	3.390
1000	0.842	1.037	1.282	1.646	1.962	2.330	2.581	3.098	3.300
Infinite	0.842	1.036	1.282	1.645	1.960	2.326	2.576	3.090	3.291

Critical values of *t* for two-tailed test
Significance level (α)

Degrees of freedom *(df)*	0.2	0.15	0.1	0.05	0.025	0.01	0.005	0.001	0.0005
1	3.078	4.165	6.314	12.706	25.452	63.657	127.321	636.619	1273.239
2	1.886	2.282	2.920	4.303	6.205	9.925	14.089	31.599	44.705
3	1.638	1.924	2.353	3.182	4.177	5.841	7.453	12.924	16.326
4	1.533	1.778	2.132	2.776	3.495	4.604	5.598	8.610	10.306
5	1.476	1.699	2.015	2.571	3.163	4.032	4.773	6.869	7.976
6	1.440	1.650	1.943	2.447	2.969	3.707	4.317	5.959	6.788
7	1.415	1.617	1.895	2.365	2.841	3.499	4.029	5.408	6.082
8	1.397	1.592	1.860	2.306	2.752	3.355	3.833	5.041	5.617
9	1.383	1.574	1.833	2.262	2.685	3.250	3.690	4.781	5.291
10	1.372	1.559	1.812	2.228	2.634	3.169	3.581	4.587	5.049
11	1.363	1.548	1.796	2.201	2.593	3.106	3.497	4.437	4.863
12	1.356	1.538	1.782	2.179	2.560	3.055	3.428	4.318	4.716
13	1.350	1.530	1.771	2.160	2.533	3.012	3.372	4.221	4.597
14	1.345	1.523	1.761	2.145	2.510	2.977	3.326	4.140	4.499
15	1.341	1.517	1.753	2.131	2.490	2.947	3.286	4.073	4.417
16	1.337	1.512	1.746	2.120	2.473	2.921	3.252	4.015	4.346
17	1.333	1.508	1.740	2.110	2.458	2.898	3.222	3.965	4.286
18	1.330	1.504	1.734	2.101	2.445	2.878	3.197	3.922	4.233
19	1.328	1.500	1.729	2.093	2.433	2.861	3.174	3.883	4.187
20	1.325	1.497	1.725	2.086	2.423	2.845	3.153	3.850	4.146
21	1.323	1.494	1.721	2.080	2.414	2.831	3.135	3.819	4.110
22	1.321	1.492	1.717	2.074	2.405	2.819	3.119	3.792	4.077
23	1.319	1.489	1.714	2.069	2.398	2.807	3.104	3.768	4.047
24	1.318	1.487	1.711	2.064	2.391	2.797	3.091	3.745	4.021
25	1.316	1.485	1.708	2.060	2.385	2.787	3.078	3.725	3.996
26	1.315	1.483	1.706	2.056	2.379	2.779	3.067	3.707	3.974
27	1.314	1.482	1.703	2.052	2.373	2.771	3.057	3.690	3.954
28	1.313	1.480	1.701	2.048	2.368	2.763	3.047	3.674	3.935
29	1.311	1.479	1.699	2.045	2.364	2.756	3.038	3.659	3.918
30	1.310	1.477	1.697	2.042	2.360	2.750	3.030	3.646	3.902
40	1.303	1.468	1.684	2.021	2.329	2.704	2.971	3.551	3.788
50	1.299	1.462	1.676	2.009	2.311	2.678	2.937	3.496	3.723
60	1.296	1.458	1.671	2.000	2.299	2.660	2.915	3.460	3.681
70	1.294	1.456	1.667	1.994	2.291	2.648	2.899	3.435	3.651
80	1.292	1.453	1.664	1.990	2.284	2.639	2.887	3.416	3.629
100	1.290	1.451	1.660	1.984	2.276	2.626	2.871	3.390	3.598
1000	1.282	1.441	1.646	1.962	2.245	2.581	2.813	3.300	3.492
Infinite	1.282	1.440	1.645	1.960	2.241	2.576	2.807	3.291	3.481

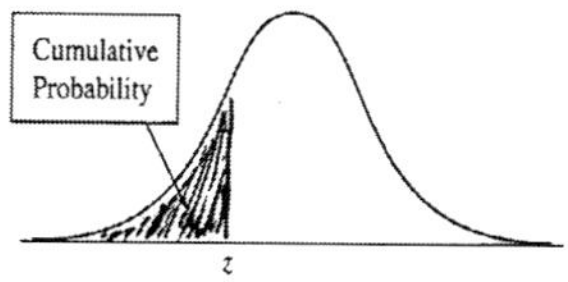

Cumulative probability for z is the area under the standard normal curve to the left of z

z	.00
−5.0	.000000287
−4.5	.00000340
−4.0	.0000317
−3.5	.000233

TABLE A Standard Normal Cumulative Probabilities

z	.00	.01	.02	.03	.04	.05	.06	.07	.08	.09
−3.4	.0003	.0003	.0003	.0003	.0003	.0003	.0003	.0003	.0003	.0002
−3.3	.0005	.0005	.0005	.0004	.0004	.0004	.0004	.0004	.0004	.0003
−3.2	.0007	.0007	.0006	.0006	.0006	.0006	.0006	.0005	.0005	.0005
−3.1	.0010	.0009	.0009	.0009	.0008	.0008	.0008	.0008	.0007	.0007
−3.0	.0013	.0013	.0013	.0012	.0012	.0011	.0011	.0011	.0010	.0010
−2.9	.0019	.0018	.0018	.0017	.0016	.0016	.0015	.0015	.0014	.0014
−2.8	.0026	.0025	.0024	.0023	.0023	.0022	.0021	.0021	.0020	.0019
−2.7	.0035	.0034	.0033	.0032	.0031	.0030	.0029	.0028	.0027	.0026
−2.6	.0047	.0045	.0044	.0043	.0041	.0040	.0039	.0038	.0037	.0036
−2.5	.0062	.0060	.0059	.0057	.0055	.0054	.0052	.0051	.0049	.0048
−2.4	.0082	.0080	.0078	.0075	.0073	.0071	.0069	.0068	.0066	.0064
−2.3	.0107	.0104	.0102	.0099	.0096	.0094	.0091	.0089	.0087	.0084
−2.2	.0139	.0136	.0132	.0129	.0125	.0122	.0119	.0116	.0113	.0110
−2.1	.0179	.0174	.0170	.0166	.0162	.0158	.0154	.0150	.0146	.0143
−2.0	.0228	.0222	.0217	.0212	.0207	.0202	.0197	.0192	.0188	.0183
−1.9	.0287	.0281	.0274	.0268	.0262	.0256	.0250	.0244	.0239	.0233
−1.8	.0359	.0351	.0344	.0336	.0329	.0322	.0314	.0307	.0301	.0294
−1.7	.0446	.0436	.0427	.0418	.0409	.0401	.0392	.0384	.0375	.0367
−1.6	.0548	.0537	.0526	.0516	.0505	.0495	.0485	.0475	.0465	.0455
−1.5	.0668	.0655	.0643	.0630	.0618	.0606	.0594	.0582	.0571	.0559
−1.4	.0808	.0793	.0778	.0764	.0749	.0735	.0721	.0708	.0694	.0681
−1.3	.0968	.0951	.0934	.0918	.0901	.0885	.0869	.0853	.0838	.0823
−1.2	.1151	.1131	.1112	.1093	.1075	.1056	.1038	.1020	.1003	.0985
−1.1	.1357	.1335	.1314	.1292	.1271	.1251	.1230	.1210	.1190	.1170
−1.0	.1587	.1562	.1539	.1515	.1492	.1469	.1446	.1423	.1401	.1379
−0.9	.1841	.1814	.1788	.1762	.1736	.1711	.1685	.1660	.1635	.1611
−0.8	.2119	.2090	.2061	.2033	.2005	.1977	.1949	.1922	.1894	.1867
−0.7	.2420	.2389	.2358	.2327	.2296	.2266	.2236	.2206	.2177	.2148
−0.6	.2743	.2709	.2676	.2643	.2611	.2578	.2546	.2514	.2483	.2451
−0.5	.3085	.3050	.3015	.2981	.2946	.2912	.2877	.2843	.2810	.2776
−0.4	.3446	.3409	.3372	.3336	.3300	.3264	.3228	.3192	.3156	.3121
−0.3	.3821	.3783	.3745	.3707	.3669	.3632	.3594	.3557	.3520	.3483
−0.2	.4207	.4168	.4129	.4090	.4052	.4013	.3974	.3936	.3897	.3859
−0.1	.4602	.4562	.4522	.4483	.4443	.4404	.4364	.4325	.4286	.4247
−0.0	.5000	.4960	.4920	.4880	.4840	.4801	.4761	.4721	.4681	.4641

- Statistics: The Art and Science of Learning from Data

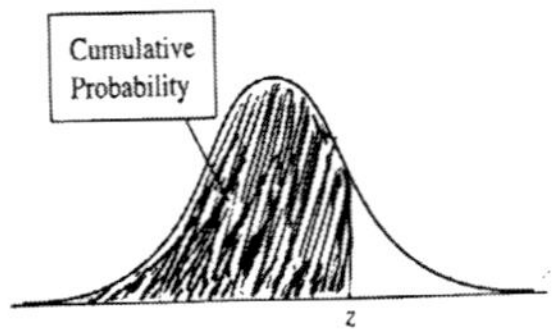

Cumulative probability for *z* is the area under the standard normal curve to the left of *z*

TABLE B Standard Normal Cumulative Probabilities

z	.00	.01	.02	.03	.04	.05	.06	.07	.08	.09
0.0	.5000	.5040	.5080	.5120	.5160	.5199	.5239	.5279	.5319	.5359
0.1	.5398	.5438	.5478	.5517	.5557	.5596	.5636	.5675	.5714	.5753
0.2	.5793	.5832	.5871	.5910	.5948	.5987	.6026	.6064	.6103	.6141
0.3	.6179	.6217	.6255	.6293	.6331	.6368	.6406	.6443	.6480	.6517
0.4	.6554	.6591	.6628	.6664	.6700	.6736	.6772	.6808	.6844	.6879
0.5	.6915	.6950	.6985	.7019	.7054	.7088	.7123	.7157	.7190	.7224
0.6	.7257	.7291	.7324	.7357	.7389	.7422	.7454	.7486	.7517	.7549
0.7	.7580	.7611	.7642	.7673	.7704	.7734	.7764	.7794	.7823	.7852
0.8	.7881	.7910	.7939	.7967	.7995	.8023	.8051	.8078	.8106	.8133
0.9	.8159	.8186	.8212	.8238	.8264	.8289	.8315	.8340	.8365	.8389
1.0	.8413	.8438	.8461	.8485	.8508	.8531	.8554	.8577	.8599	.8621
1.1	.8643	.8665	.8686	.8708	.8729	.8749	.8770	.8790	.8810	.8830
1.2	.8849	.8869	.8888	.8907	.8925	.8944	.8962	.8980	.8997	.9015
1.3	.9032	.9049	.9066	.9082	.9099	.9115	.9131	.9147	.9162	.9177
1.4	.9192	.9207	.9222	.9236	.9251	.9265	.9279	.9292	.9306	.9319
1.5	.9332	.9345	.9357	.9370	.9382	.9394	.9406	.9418	.9429	.9441
1.6	.9452	.9463	.9474	.9484	.9495	.9505	.9515	.9525	.9535	.9545
1.7	.9554	.9564	.9573	.9582	.9591	.9599	.9608	.9616	.9625	.9633
1.8	.9641	.9649	.9656	.9664	.9671	.9678	.9686	.9693	.9699	.9706
1.9	.9713	.9719	.9726	.9732	.9738	.9744	.9750	.9756	.9761	.9767
2.0	.9772	.9778	.9783	.9788	.9793	.9798	.9803	.9808	.9812	.9817
2.1	.9821	.9826	.9830	.9834	.9838	.9842	.9846	.9850	.9854	.9857
2.2	.9861	.9864	.9868	.9871	.9875	.9878	.9881	.9884	.9887	.9890
2.3	.9893	.9896	.9898	.9901	.9904	.9906	.9909	.9911	.9913	.9916
2.4	.9918	.9920	.9922	.9925	.9927	.9929	.9931	.9932	.9934	.9936
2.5	.9938	.9940	.9941	.9943	.9945	.9946	.9948	.9949	.9951	.9952
2.6	.9953	.9955	.9956	.9957	.9959	.9960	.9961	.9962	.9963	.9964
2.7	.9965	.9966	.9967	.9968	.9969	.9970	.9971	.9972	.9973	.9974
2.8	.9974	.9975	.9976	.9977	.9977	.9978	.9979	.9979	.9980	.9981
2.9	.9981	.9982	.9982	.9983	.9984	.9984	.9985	.9985	.9986	.9986
3.0	.9987	.9987	.9987	.9988	.9988	.9989	.9989	.9989	.9990	.9990
3.1	.9990	.9991	.9991	.9991	.9992	.9992	.9992	.9992	.9993	.9993
3.2	.9993	.9993	.9994	.9994	.9994	.9994	.9994	.9995	.9995	.9995
3.3	.9995	.9995	.9995	.9996	.9996	.9996	.9996	.9996	.9996	.9997
3.4	.9997	.9997	.9997	.9997	.9997	.9997	.9997	.9997	.9997	.9998

z	.00
3.5	.999767
4.0	.9999683
4.5	.9999966
5.0	.999999713

Index